STARTING OUR CAREERS

*A Collection of Essays and Advice
on Professional Development from the
Young Mathematicians' Network*

**Curtis D. Bennett
Annalisa Crannell**

Editors

AMERICAN MATHEMATICAL SOCIETY

1991 *Mathematics Subject Classification*. Primary 00A35; Secondary 00A06.

Library of Congress Cataloging-in-Publication Data
Starting our careers : a collection of essays and advice on professional development from the young mathematicians' network / Curtis D. Bennett and Annalisa Crannell, editors.
p. cm.
Includes bibliographical references and index.
ISBN 0-8218-1543-1
1. Mathematics—Vocational guidance. I. Bennett, Curtis D. II. Crannell, Annalisa.
QA10.5.S7   1999
510′.23—dc21

99-14350
CIP

For Jonathan and Samuel
(*cdb*)

To Mama and Dad, who showed me how
(*akc*)

Contents

Introduction

This is the best of times; this is the worst of times. (Or, as Marge Murray says in Chapter 4, "This is a time of trouble, but also of opportunity"). If you are the reader we envision for this book, you have just passed through the most crucial stage of your career—writing and successfully defending your doctoral thesis in mathematics—only to discover that what lies ahead of you is, yet again, the most crucial stage of your career. It is the time when you make the choice about what job to take (or allow circumstances to make the choice for you); it is the time when you make the adjustment from studying in a research institution to earning your keep in industry or in an undergraduate college or in another research institution; it is the time when you will—or will not—publish your thesis, it is the time when you will decide to leave research behind you or to start new mathematics on your own, or when you will struggle to balance time for students and committees with time in the library.

This book was written largely by people like you. It began as a weekly e'mail newsletter called the Concerns of Young Mathematicians (CoYM), whose editors were themselves a small band of young mathematicians. (Here as elsewhere in this book "young" measures time since the doctorate rather than biological age). The questions that these editors and their writers addressed in the newsletters were ones that naturally concerned them and their readership (and will therefore, we hope, naturally concern you): How do I get good letters of recommendation? How do I apply for a grant? How do I do research in a small department that has no one in my field? How do I do *anything* meaningful if all I can get is a series of one-year jobs?

One question which the CoYM most frequently addressed—how to apply for jobs—plays a very small role in this book. There are two reasons for this, neither of which has to do with indifference for the plight of the unemployed or under-employed mathematicians.

The first reason is that the job market in mathematics has gotten so much attention during the last decade that good information is both plentiful and readily accessible—as you can see from our bibliography. The second reason is that professional development has gotten so little attention, despite its obvious importance to young mathematicians, that we decided to use this book to plug that gaping hole.

With this one caveat, however, the articles in this book paint a broad portrait of the professional development issues that most interested the Young Mathematician's Network. So you will see that there was considerable curiosity about making the transition from academia to industry—how to make contacts, how to prepare oneself with appropriate course work and extra curricular activities—but that very few people had questions about how to proceed once there. On the other hand, working in academia is fraught with questions about teaching, research, giving talks, and so on. The fact that the chapter on tenure has only two articles is an artifact of the youth of CoYM: readers and writers of the newsletter were more interested in "how to get started on the road to success" than the eventual documentation of this success.

Above we wrote that "this book was written largely by people like you". You will see, however, that the editors of CoYM followed their own advice about seeking help from more senior (and luminary) mathematicians. Almost every section of this book contains voices of people who are struggling as you struggle now, together with articles by mathematicians with a more distant, but much broader, perspective. We (Curt and Annalisa) are deeply grateful to all of the authors, those young as well as those ... experienced, who contributed to the CoYM in its early years and who graciously allowed us to include their work in this volume.

A Brief History of the Young Mathematicians' Network

In 1993 five junior mathematicians, Edward Aboufadel, Kalin Godev, Mark Winstead, and Charles Yeomans, and Curtis Bennett, decided to start an electronic group called the Young Mathematicians Network (YMN), where young refers to the length of time since obtaining one's Ph.D. This group initially grew out of the Young Scientists Network (YSN), a group of young physicists and other scientists that had gathered to discredit "The Myth," the idea that jobs would soon be plentiful for math and science Ph.D.s. The five mathematicians felt there was a need for a group that specifically dealt with the problems junior mathematicians faced.

The main publication of this group, the *Concerns of Young Mathematicians* or CoYM, began in the summer of 1993 and is published weekly during the school year and biweekly during the summer. Within a few months of its start, the YMN began to thrive. Over time, many more mathematicians volunteered to join the original five editors on the editorial board, and currently, only one of the five original editors is still on the board. While an attempt at brevity prohibits mentioning all of the editors and major contributors, there are several without whom the YMN might never have gotten off the ground, and they and their contributions must be mentioned. Mark Winstead first presented the idea of a group for mathematicians, and without him, nothing ever would have happened. Charles Yeomans ran the mailing lists for the first several years, and his willingness to do the dirty work made it possible for the Concerns to reach as many people as it successfully has. Edward Aboufadel gave the group visibility as he had just completed his widely read "Job Search Diary" (which appeared in *FOCUS*) when the YMN started. Steven Kennedy joined the editorial board early on, and in addition to being a frequent contributor, he did a large amount of legwork in trying to get people to submit articles. Finally, Kevin Charlewood took over as managing editor when Curtis Bennett resigned from the board in 1995 and has helped keep the YMN a thriving group.

Originally, the YMN had five major goals:
1. to combat "The Myth,"
2. to provide information on the job market,
3. to provide a support group for junior mathematicians,
4. to keep the mathematical community informed about the concerns of junior mathematicians, and

5. to help junior mathematicians get information about professional development.

This book is an attempt to gather together the articles from the Concerns that best address professional development issues. The reader that is interested in reading old issues of the Concerns should visit the YMN archive at

`www.math.usouthal.edu/~brick/ymn/archive.html`

which as of this publication is maintained by Steve Brick. In closing, we wish to thank everyone who helped make the YMN possible by their contributions, their hard work, and their concern with the difficulties the most recent job market has created for junior mathematicians.

Acknowledgements

This book owes its very existence to John Ewing, who first suggested to us (and even urged on us) the project. It is entirely appropriate that this book advises us to get good ideas from mathematicians we admire.

The technical (and TEX-nichal) aspects of putting together a book boggled us. We owe a big round of applause to Chris Swisher for all the duct tape repairs to Annalisa's Mac, to Thomas Hern for providing helpful advice and for keeping Curt's computer up and running, and to the AMS technical support staff (especially Tom for the P/Γ help and to Barbara for the boxes). The next round of duct tape's on us, guys.

We would like to thank our departments at Bowling Green and Franklin & Marshall for their support, encouragement, honest criticism (!), coffee, avuncular advice, and collegiality during our own early careers. Big hugs are due Elizabeth Clayton-Bennett and Neil Gussman for their support during the completion of this project.

If we hadn't had so much help proofreading, we would have had to include our opthamologists in this list. Fortunately we were assisted by all of our authors (whose names are listed at the end of this book), by Michelle LeMasurier, and by our extremely patient meta-editor Ed Dunne.

But most of all, we take our hats off to all of the past, present, and future editors of the Young Mathematicians' Network, who continue to excel in an endeavor for which this book is a mere attempt.

Curtis D. Bennett
Annalisa Crannell

CHAPTER 1

Applying for Jobs

As we mentioned in the Introduction, the question of "How to Find a Job" was one of the most persistently discussed topics in the e'pages of the CoYM. The first two articles in this chapter summarize much of the advice that appeared there: the first, on the peculiarities of preparing for an academic job in which teaching is weighted heavily; and the second, on preparing for a job outside of academia altogether. The next pair of articles offer specific suggestions (and ethical considerations) regarding letters of recommendation. Much of what is said in these first four articles is echoed in the next two chapters on "Industrial Mathematics" and "Life in Small Schools".

The remaining articles in this chapter address more general issues: the two-body problem, the isolated-body problem (a tantalizing picture of working on a tropical island), and a brief but typical description of the one-year-job-blues. The article by Montgomery and Powell first appeared in the newsletter of the Committee on the Status of Women in the Economics Profession. We thank the CSWEP for permission to reprint this article.

Liberal Arts Marketability
by Steve Kennedy

There has been some discussion recently in this forum about choosing a marketable specialty. If your goal is a position at a liberal arts college, then your field is of little import. But the discussion started me thinking about what a person could do in graduate school to improve their chances of landing a job in a liberal arts college after graduation. The single most important thing you must do in graduate school is create a record of effective teaching. When the hiring committee at a liberal arts college reads your dossier, the first thing that they look for is proof that you will be a success in the classroom.

How to Create a Record of Effective Teaching. First, of course, you must work at becoming an effective teacher—that means being organized and prepared for class or recitation section, including reading the book and trying some of the problems beforehand, treating your students with respect and being understanding of their problems. There are resources for improving your teaching: most universities have pedagogy experts on site with whom you can consult, or will provide student observers to attend your class and discuss what works and what doesn't.

Read Steven Krantz's "How to Teach Mathematics." Most university mathematics departments have one or two faculty members who are concerned about teaching—identify that person in your department. Ask him or her to attend your class occasionally and discuss with you your performance. Do this many times over the course of your years in grad school. When you are looking for a job, this person will be in a position to write a meaningful, informative letter about your teaching, your growth as a teacher, and your dedication to improving your teaching. I can't stress enough how important it is to have some such documentation. A letter of recommendation from someone who has actually observed your teaching and discussed pedagogy with you is infinitely more impressive than a letter from someone who has only read your student evaluations. Everybody has a letter attesting to the fact that their evaluations "consistently rank them in the top X% of teaching assistants in the university", and almost everybody (it seems) has a teaching award. Consequently, neither of these is particularly impressive anymore. (If you do have a significant teaching award it might be smart to indicate how selective the criteria are for it in your dossier: e.g., "I received the O.I.M. Goode Prize for Teaching Excellence. This prize is awarded annually to three graduate teaching assistants in the College of Arts and Sciences. There are approximately 500 eligible graduate teaching assistants.")

It also helps to have taught classes on your own, as opposed to having only been an assistant running recitation sections. If this is not possible at your institution, consider applying for adjunct or summer work at a local two- or four-year college. Document your success there.

Be Versatile. Take a variety of courses, read books and journals (including expository ones) and get as much mathematical breadth as possible. Narrow specialization is the antithesis of the liberal arts ideal. You should be willing (and, ideally, eager) to teach almost every course in the department. During the most recent candidate search at my institution the chair of the search committee talked repeatedly of looking for a good "utility infielder." With the exception of statistics and computer science, which are really different disciplines, it is uncommon for a liberal arts school to be seeking a particular field. Most liberal arts colleges have zero or one statistician and many staff their CS courses entirely with mathematicians (for economic reasons: a computer scientist is more expensive than a mathematician). One way to improve your chances at a liberal arts school is to get some formal training in one or both of these disciplines. A department with only one statistician might like the security of having a second body around competent to teach upper-level stats.

Make Contacts and Be Informed. Join the AMS and the MAA, read their journals and their newsletters. Know what is happening in your profession. Keep abreast of developments in curricular reform. (I know people who taught reform versions of calculus in grad school and got interviews and jobs based on their experience.)

Go to meetings. Give a talk at your sectional MAA meeting. There will be people there from every local college: meet them, remember their names, write to them when you get on the job market. Use any contact you can to get your application noticed. There will be ∼1000 applicants for any job you want. The hiring committee physically can not give careful attention to every application.

There comes a point where they must be looking for reasons to stop reading an application, just so they can get through the pile. You have to give them a reason to give extra attention to your file. If the person reading your file heard you give a nice talk at an MAA meeting or had lunch with you, that might be all the break you need to get your foot in the door.

You need to make your application stand out from the pile. Doing something extra during grad school could be the way to do that. Good luck.

Preparing For A Job In Nonacademics
by Stan Benkoski

The difficult job market for Ph.D. Mathematicians in the 1990's has been well documented by many sources. As a response, one of the AMS Task Force on Employment's recommendations was that "the AMS use the various means available to it to make clear to the mathematical community the value of, and opportunities for, nonacademic employment." The purpose of this discussion is to attempt to respond, in part, to this challenge from the Task Force. In particular, these remarks will be directed at a second year graduate student. (The information should also be helpful to any graduate or under-graduate student. Much of this is also useful to a faculty member who wants to learn more about mathematics in industry.) The goal will be to provide information about the nonacademic job market and steps to prepare to obtain a job in that market.

A couple of caveats are appropriate. First, a second year graduate student has spent most of the last nineteen years in school. The academic environment is well known and understood. It is comfortable. Little is known about nonacademics. This means that the effort to learn about nonacademic mathematics usually starts from a position of very little knowledge. Second, the student will be required to do a lot of the work himself. In particular, there is currently no single source which will provide comprehensive information about nonacademic mathematics.

The remarks that follow fall into three sections. First, I briefly describe my background in order to establish my credentials and biases. Second, (and perhaps most important) is a somewhat philosophical discussion about the differences between employment in academics and nonacademics. I believe that a different mind set is required to seek a position (and to be successful) in nonacademics than is required in academics. If this different mind set is not achieved, then the specific suggestions in the third section will not be successfully employed.

Background. I have worked for Wagner Associates for twenty-one years. We are a consulting firm in Mathematics, Operations Research, and Software Development. In the thirty-one year history of the firm, we have worked on a wide variety of problems. The vast majority of our work had been funded by government agencies. In particular, the Department of Defense (in various guises) has been our biggest sponsor. A lot of our work has been in the search for lost objects.

I received my Ph.D. in Number Theory from The Pennsylvania State University in 1973. Two points should be noted. First, my academic training was in pure mathematics. I did old-fashioned elementary number theory. (My Erdös number is 1.) Second, 1973 was part of the last big slump in employment opportunities for mathematicians, and the experience of looking for a job in that environment gives me some empathy with current job seekers.

While my academic training was in pure mathematics, I had three summer jobs which used applied mathematics. Two of these were government jobs. The technical work in these jobs involved operations research and software.

Wagner Associates was, and still is, unusual in that we seek (when hiring) research quality in mathematics and not necessarily an education in applied mathematics. However, I am sure that my work experiences allowed me to stand out from the crowd. I first made contact with Wagner Associates at the Employment Register at the Annual Meeting in 1973.

I personally found the transition to industry to be a relatively easy one. (Some of that ease of transition must be attributed to my previous experience with summer jobs.) I have thoroughly enjoyed the breadth and depth of work that I have done but it is quite different from an academic experience and each individual has different priorities and goals.

Philosophy. The change in philosophy that is required for a good nonacademic job search can be summed up in two sentences. In academics, you get a job if they believe that you are smart. In nonacademics, you get hired if they believe that you can help them.

The graduate student has 100% of the responsibility to find out what opportunities are available and to convince an industrial employer that he could make a contribution.

People in nonacademics tend to have a much different view of the world. The philosophy of most Mathematics Departments is that all mathematics is intrinsically valuable and that our society should (and must) support that endeavor. Academic Mathematics is not exactly a public works project, but rather based on the belief that mathematics enriches our lives and also has demonstrated the potential to profoundly change our lives. In particular, mathematics is different from fine arts or philosophy. Endeavors such as these (fine arts and philosophy), are supported with dollars from our society because we believe that they are important to our culture. They enrich our lives and are part of the "examined" life. But clearly mathematics is different from this. Without digressing into philosophical considerations, the fact is that mathematics is funded at a much higher level. (After all, how many jobs are there for Ph.D. philosophers?) We like to think mathematics is somehow a high-level endeavor that is intrinsically superior to other studies. The crass truth is that the financial support for mathematics arises primarily as an "enabling technology." Its application and practical use are what give it a privileged position.

The perspective of business is quite different. Few businesses believe that they can afford basic mathematical research, and probably cannot justify it to their stockholders. As an institution, a business' interest in a mathematician is based on solving problems that require mathematics. The company's desire is to do something better, faster, smaller, cheaper, *etc.* Mathematics is a means, not an end. Industrial mathematics problems rarely appear as math problems. The value of a mathematician to industry is the ability to take a problem which is posed as a real, practical problem (or perhaps something that is not even perceived as a problem); state that problem in mathematical terms; and proceed to develop insight into the problem which results in quantifiable improvements.

As a mathematician who has worked in industry for twenty-two years, it is amusing and somewhat annoying to observe the negative correlation between the job prospects for mathematicians and the (academic) mathematical community's interest in greater involvement in industry. If the (academic) mathematics community really believed doing mathematics in industry is a noble profession, then the community would be interacting with industry and sending a share of the best students into nonacademics in good times as well as bad. But the interest ebbs and flows with the job market and, when hard times hit, nonacademic opportunities get more attention.

Specific Suggestions. I would give some specific suggestions for resources to learn more about nonacademic employment and to be better prepared to investigate nonacademic opportunities. These are:

1. College career centers
2. Reading/seminars/short courses, *etc.*.
3. Academic classes in a broad range of subjects
4. Experience

Of these, the most valuable thing that can be done is to gain experience. This takes some real effort but is worth it. I will discuss some approaches that can be used to obtain this experience but first will discuss the other topics.

Many college career centers are excellent places to learn about nonacademic employment. Most of the information provided there will be about nonacademic employers. Lots of literature will be available that describes what these employers do and what sorts of skills and talents they are looking for. In addition, some of these employers may be looking for part-time, internship, or summer help. Visit this center and find out what they have to offer.

One of the goals of this process is to learn about applications of mathematics. Activities such as reading, attending seminars, attending short courses, *etc.*. are effective tools for this process. SIAM puts out a number of publications that would be useful. In addition, the AMS-MAA-SIAM Mathematics and Industry Project www.ams.org/careers/ is directed at improving the match between graduate education and industry. The Journal of Operation Research would be a good place to start to learn about Operations Research. Other good topics are biotechnology and digital signal processing.

Taking academic classes in fields other than Mathematics, in order to expand the knowledge of application of Mathematics, is another useful idea. The most directly applicable courses will be in Computer Science. Computer Science comes first, and foremost, because almost any summer or part-time job will involve writing software. Most full-time jobs will require (as a minimum) a working knowledge of software and may well require the ability to write good code. Other suggestions would be Numerical Analysis, Biology, Economics, Engineering, Physics, Statistics, and Operations Research.

The most important suggestion is to get some experience in a nonacademic environment. First of all, it would provide experience in what goes on in a nonacademic job. The best way to learn anything is to do it. This experience also provides an opportunity to determine one's aptitude and interest in nonacademic employment. It is different than academic employment. (Of course, any one particular experience is a very small and biased sample of nonacademics. Another job at the same company or a different company could be a totally different experience.)

Second, the best reference when applying for a job is solid evidence that one has already accomplished something similar.

How does one get such experience? One of the best bets is summer employment. (Most of what follows also applies to part-time work which would also be useful.) There are two basic sources of summer jobs: government and industry. Government is probably the best source since there are fewer institutions to contact and they often have specific programs to support summer employment. They will also have employment offices which will have lists of available positions and directions on how to apply. Large companies will have personnel offices which can provide the same information. These methods can be successful but they require work. There are also on-line services such as Help Wanted-USA and E-span employment database.

An even better method is to get a contact inside the company. One approach to this is called networking. Talk to your friends, acquaintances, professors, *etc.* Let it be known that you are looking for summer employment. (Note, as mentioned before, the Career Center on campus may also be a rich source of information in this regard.)

One very effective process is known as consult visits or informational interviewing. A consult visit consists of a 20 minute interview with someone in industry who is involved in applying mathematics. The steps in the process are:

1. Develop names and addresses of individuals involved in mathematics in industry.
2. Research the company and the person you will visit.
3. Write letters to some of those individuals.
4. Make follow-up phone calls.
5. Interview.
6. Follow-up.

The first step in this process is to find individuals who may be appropriate for consult interviews.

The Combined Membership List is a good source of names and addresses of mathematicians who live in your vicinity and don't work in academics. Recently, I was asked to speak at the University of North Texas on the job opportunities in nonacademic mathematics. The Combined Membership List contains eleven companies in the greater Dallas area that have employees who belong to AMS, MAA, and/or SIAM. There are a total of 107 listed members who are not associated with a college or university. This would appear to be a rich source of possible contacts for consult interviews.

Figure 1 gives an example of a consult letter. This should be tailored as much as possible by mentioning the company name, *etc.*

In the follow-up phone call, simply mention that you are following up on your letter and would like to know if it would be possible to schedule a 20-minute appointment to discuss the application of mathematics in that particular company.

Prior to the consult interview, a substantial amount of preparation is required. In particular, one must know what sort of business the company is in and as much as possible about the individual who will be interviewed. Sometimes this can be accomplished by simply calling the company and asking for appropriate information. Other sources include the library, on-line newspapers, *Hoover Handbook Company Profiles*, trade magazines, *etc.* As a minimum, you should know what the company's main lines of business are.

Figure 1. *Sample Consult Letter*

Your Name (Typed On Letterhead)
Your Address
Your Phone Number

Name
Title
Company
Address
City, State, Zip

Dear [So-and-So],
I am a (graduate student, undergraduate, ...) at (name of
school) in mathematics and I am interested in learning more
about the opportunities for mathematicians who are outside of
academics. I obtained your name from the Combined Membership List which indicates that you are a member of (SIAM, AMS,
MAA, as appropriate). I would like to ask for 20 minutes of your
time to hear your views on the opportunities for mathematicians
at an organization like (company name). In particular, I would
like to discuss the kind of mathematical problems that your company faces. I would be glad to meet at your office at a mutually
agreeable time.
I will call in the next week to determine if we can arrange a time
to meet.

Sincerely,
(Original Signature)
Your Name

The interview itself should be conducted as an interview of an expert. The
objective is to determine how mathematics is being used (or could be used) at that
company. If, during the interview, opportunities arise to discuss possible part-time,
internship, or summer work, then this is a golden opportunity.

At the end of the interview, ask the individual if they can recommend anyone
else who could provide information on mathematics in industry. Follow-up on those
leads.

Send a thank you letter within three days of the consult interview.

Some diligent work with consult interviews should produce a much better idea
of what is done in nonacademic mathematics plus some possible leads for summer
or part-time jobs.

Summary. It is not an easy task to effectively investigate nonacademic mathematics. The skills required for this effort are not advanced skills, they are skills that are just not taught in academics. It also requires a serious commitment of time.

At worst, the result of such an effort will be the realization that only academic mathematics are of interest. At best, the result would be a set of rich and challenging opportunities in mathematics in a nonacademic setting.
[*Editor's note*: The above is a transcription excerpted from a talk by the author].

Getting Good Letters of Recommendation
by Annalisa Crannell

This is in response to a mathematician's request for advice on how to get good letters of recommendation. He was distressed over the fact that his letters were bland, and hardly personal. He said that this was true even in letters from professors that had seen his teaching and had read all his student evaluations.

Put yourself in the letter writer's shoes. You're busy, you've got to write letters for people, and you really don't know a whole lot more than stuff you've seen written down: a thesis, student evaluations, *etc.* Some professors try to learn more about people they're writing for (see the next paragraph, for example), but a lot of people give in to ennui. It's easy to think that professors should know about you by reputation or by actually remembering good things about you, but that's generally not the case. Those of you who write letters for students in your own sections of calculus (or whatever) know how hard it is to say something meaningful.

The mathematician who asked this question wrote: "One of my committee members asked for a copy of my resume. He then asked questions about what was listed there and mentioned some of my entries in his letter. This certainly helped personalize it. "

I really believe that this is the key to getting more individualized letters. I think it's in bad form to chastise your letter writers about being too generic, nor is it a good idea to ask them to "Please say *this*". On the other hand, it's a *great* idea to give them as much information as you can, and let them cull their own excerpts.

Before you go ask for letters of recommendation, put together a folder that contains all sorts of stuff that you're proud of. Start this the first year of grad school. Folks in the know call these "Portfolios"; but mine was called "Bragging Folder" for a long time.

In this folder, put anything that you're proud of. This might include:
1. copies of syllabi,
2. letters from students, parents, and deans
3. material from conferences you've attended
4. lists of grants you've received
5. records of how many students continued in math after passing your class
6. announcements from talks you've given
7. resume
8. innovative teaching/computing/research you've done.

Then, organize this folder, make a table of contents, and take this to your letter writers. You might even give them a list of the places you're applying to (they might know someone at one of these institutions, or give you more advice). Your letter writers will love having the information at their fingertips (as opposed to in the dark recesses of their minds); you'll get a more individual and more accurate letter.

Letters of Recommendation Should Be Confidential
by Evelyn Hart

I'd like to discuss confidentiality of letters of recommendation.

I realize that some states have laws making confidentiality impossible, but the whole system is weakened when letters aren't confidential.

I write many letters for my students. I never agree to write a letter if I don't think it will be a good one, but even so I get the creeps when the student I'm writing about sees the letter. (For many scholarships, the directions say that everything must be sent in one large envelope, so I can't send the letter myself.) No matter what I say, won't the student think I should have said something else?

My solution is to ask the student to agree not to look at the letter, and I put it in a sealed envelope and sign over the flap. Inside, across the top of the letter I can then write: "Confidential letter of recommendation for ... ". I tell the student (and I believe it is true) that the letter will be taken more seriously that way. I also assure the student that I have plenty of good things to say, and I often say what those things are.

As a member of a department that hired this year, I can say for sure that there are plenty of dull letters out there. But what is worse is that they may be so uniformly dull because the authors think the candidate may eventually read the letter. If somehow this were impossible, I think letters would become more interesting and more accurate and useful. Several letters I read did say, "This person is as good as so-and-so and definitely better than what's-his-name." I knew the people concerned, and it was very helpful to have that information. I intend to keep that information confidential, of course. But maybe I'm too old fashioned. The authors of those letters went out on a limb to do their job well despite the fact that the letters might not remain confidential. Most are not willing to take chances.

So is there any way to solve the problem? Being able to see letters about you might protect you from being hurt badly in one letter, but it also weakens the entire process for everyone.

And the Two Shall Be as One: Job Sharing in an Academic Department
by Mark Montgomery and Irene Powell

It is quite common these days for both members of a married couple to have the same level of professional training. Often they will both be trained in the same field. But for academic couples, there is a simple rule of thumb: if you have the same level of training, in the same field of study, then you may not have the same zip code. In other words, if you fall in love with another Ph.D., don't expect to whisper sweet nothings to him without the help of Sprint or MCI. A modern alternative to the long-distance marriage, however, is the sharing of one academic position. Or maybe the sharing of one-point-something academic positions. For example, the authors of this article each represent between .7 and 1.0 associate professors at Grinnell College, with an annual mean of .9, and a standard deviation of about .1. (This is not something we try to explain at cocktail parties.) Our joint contract stipulates that we teach no fewer than seven courses between us per year, and up to ten (full-time) if we and the college mutually agree. In this essay we consider the advantages and disadvantages of this kind of arrangement.

Not every college or university is ready to embrace the shared academic contract. In fact, it will only happen for one of two reasons:

1) An angel of the Lord appears to the dean and demands he implement a shared contract policy,

2) The school has trouble recruiting women.

Often both things have to happen. A place like Grinnell, where we teach, has a strong incentive to resort to shared contracts because it is a small college, in a town of 8000, deep in the heart of Iowa. Rural Iowa. A woman professor here may have trouble finding her husband a job better than the one Henry Fonda had in The Grapes of Wrath. If the husband is in the same academic field, a shared contract may be an attractive option.

There are goods things and bad things about shared contracts, of course. Obviously, a couple sharing a job earns less money than if both spouses had full-time jobs at separate institutions. As economists, we can't dismiss this as a trivial disadvantage. But the earnings differential may be smaller than you think. Even if you share a single full-time job, you can usually pick up extra courses to teach. In any given year, there is nearly always someone in your department on leave. When we were recruited, our department was actively seeking a job-sharing couple to create some flexibility in leave replacement. Moreover, don't forget that there is complementary slackness between your budget and time constraints. If you earn less money, it's because you have less teaching to do. For two assistant professors who want to establish research records, the extra time is a major advantage. For example, the lower time requirements of the shared contract will certainly make it much easier for you to have children. (Actually we couldn't decide whether that's an advantage or a disadvantage of a shared contract, so let's just call it a "feature.") And finally, consider another terrific aspect of splitting a single job: you and your spouse will spend more time together every day than almost any couple you know. A lot more time. (Well, here again, better just call that a feature.)

All right then, let's suppose you and your spouse have abandoned the dream of twin endowed chairs at Stanford and are willing to try a shared position. You have found an institution that is open-minded, progressive, and in tune with cutting-edge innovations in personnel management. Or, at least, one that's out in the sticks. What should you negotiate in your shared contract? First, of course, find out how tenure and promotion requirements will be handled by your institution. Work this out up front. One position-sharing couple we know was told by their chair that he expected twice as much research from each of them because each would be doing only half as much teaching. WRONG! [Loud Buzzer]. The point of a shared contract is not for the college to get four times as much research for the same salary. Technically, for a single paycheck, the college should expect that you'll each do half as much teaching, half as much research, and spend a lot more time watching *Days of Our Lives*. In fact, of course, you'll do more than that. The school will get more research and service than they would from a full-time professor, and that redounds to the greater glory of both the college and the couple. All well and good. But don't give them the right to demand it of you.

Along the same lines, how much college service will you do? At Grinnell, we typically each do as much student advising and serve on as many committees as full-time faculty members. We know a couple at Kenyon College, however, who have a similar shared contract, but apparently more brains, because they negotiated the right to limit their committee work.

Speaking of teaching, research and service, how will yours be evaluated: will they treat you as a couple or two individuals? For promotion and tenure Grinnell reviews each partner in the shared contract separately, which is how everyone involved seems to prefer it. In this respect the shared contract is not much different than two positions. But for the joint position there is an additional question of timing. One school of thought holds that two half-timers should get twice as many years to produce a tenurable portfolio as full-time faculty. Our friends at Kenyon, for example, were offered seven to ten years, at their own option. Usually, however, the institution is not opposed to your coming up in the regular time frame. The final important issue regarding tenure is what happens if only one of you gets it. Do both partners have to hit the road, or may one of you assume a full-time position? We have heard of at least one case where neither is permitted to stay.

Salary issues can be a little complicated, especially when raises are based on merit. Until recently, Grinnell insisted that we be paid identical salaries, prorated by the number of courses we taught. As it happened, one year Powell ranked 5 on the five-point merit scale, while Montgomery ranked only 3. So they averaged and gave us both a rank-4 raise. As a result, Montgomery became professionally jealous of Powell and Powell began to think of Montgomery as a drag on her career. Recent reform made it possible for a couple at Grinnell to opt for separate salaries. For us, financially, that should be worthwhile—those divorce attorneys cost a fortune.

There are lots of other details to work out with a shared contract, some of them things you might not think of. Benefits can be an issue, of course, such as whether you both get full life insurance, and whether you each get full research support, including travel to professional meetings. (At Grinnell, "yes" to all of the above.) Also, you need to determine how sabbatical pay is handled: is it based on one full-time teaching load, or on the amount of teaching you have both actually done. (At Grinnell, it's the latter, averaged over the previous six years.) What happens if you eventually get divorced (see the discussion of merit pay above), may one of you automatically assume full duties? And how about parental leave and early retirement? These are all things to pay attention to.

For many young academic couples a shared contract is an idea worth considering, at least as a starting point for two careers. It may be very attractive when compared with having one partner languish in temporary positions so that the two of you can maintain one household. If one of you already works at an institution that doesn't have this arrangement, you might try suggesting it. If you do, be prepared for your dean to make a really funny face. (It may be worth mentioning it just to see that.) Anyone who would like a copy of Grinnell's contract arrangement can write to the authors at Dept. of Economics, Grinnell College, Grinnell IA, 50112.

My Experience with the Two Body Problem
by Jean E. Taylor

Here is my experience with the two-body problem:

My first husband got his Ph.D. three years before I did. I was in three different mathematics graduate programs, basically following him around. However, it all worked out very well for me: I think I got something valuable out of each program, my NSF graduate fellowship followed me around, and I eventually wound up getting my Ph.D. from Princeton while following my husband to his postdoc at the Institute for Advanced Study.

In my last year of graduate study, we looked for jobs together. Early on we were separately each offered non-tenure-track positions at MIT. Then various other places offered us one job jointly, then maybe 1.5 jobs, then maybe 1.75 jobs. We got sick of the negotiations and took the MIT offers. Then we got divorced.

I looked for jobs back in the New York/New Jersey area, was offered a tenure track position at Rutgers University, and took it; I've been there ever since. I married Fred Almgren, who was already tenured at Princeton University. Our situation is about as good as one could hope for, except that I have to commute 45 minutes to get to Rutgers and 45 minutes to get back home. I hate that commute. But the house came with the husband (and a 2.5% mortgage, and kids who liked Princeton schools), so I am stuck with it. (Many Rutgers faculty live in Princeton on purpose, accepting the commute because they like the town.)

There is one drawback common to all two-career couples, even when they have positions as good as ours. That is that one cannot easily get a higher-paid job offer elsewhere and move. And moving is the fundamental bargaining chip on getting higher salary everywhere. On the other hand, if you are happy where you are and don't care that much about salary, this is not a problem.

It is my observation that a lot of young women mathematicians have married men mathematicians who are senior to them, and that a large number of these young women eventually stop doing much mathematics research. While there are always individual circumstances at work, and while there can be very legitimate reasons for making one choice of how to spend your life over another, it is my feeling that part of the reason for this loss of women is that the constant exposure to someone who you feel is better than you can be very demoralizing. You have to develop a very secure sense of self to keep working in a situation like that, and to stand up for your own research needs and not buy the argument that your partner's work is worth more than yours so you should do the chores.

Finally I would like to say that the rewards of the struggle for two jobs and for two research careers is well worth it. Having a household where you can share the love of mathematics with your partner makes dinner table conversations much more interesting. In fact, we seem to communicate our enjoyment of our jobs well enough that our three kids have all decided that doing mathematics is the best job in the world—and in fact they are all gravitating towards what I do, namely equilibrium and growth problems for soap bubbles and crystals.

[*Note added in 1999*: This article was written in December 1995, before the illness and death of my husband, Fred Almgren. In retrospect, I can only emphasize that the joys of being part of a two-body vastly outweigh the inconveniences of dealing with a two-body problem.]

The Isolated Body Problem
by Raymond Grinnell

A number of articles have appeared in the CoYM on the topic of the "Two-Body Problem" and variations of it. One such variation is in the title of my article: "The Isolated Body Problem". This is the problem which deals with doing mathematics in a vacuum.

I work at the University of the West Indies in Barbados. A bit of geography is helpful. Barbados is about three hours by air south east of Miami. The island is 166 square miles and has a population of 265,000. There is one university here but the other two campuses of UWI are in Trinidad and Jamaica. Our campus has 3,000 students and is mostly concerned with teaching at the undergraduate level. We offer a three year degree which is based on the system in the UK. My department has 10 members, of which 6 are in computer science and 4 (including myself) do mathematics. It is accurate to say that myself and perhaps one other person in the department does any serious research. The average income in Barbados is about $6,000 (U.S. dollars) per year.

Part of the reason I mentioned the facts above is that they are interesting. When I left Canada in 1993 to work here, I had no idea about this place other than it is where one might go for a vacation. But the other two main reasons for the paragraph above are: (1) it points to the problem of working in isolation, and (2) it points to the problem of desperation and desire with regards to having a career in mathematics at a university.

When I finished my Ph.D. in 1991, I had two one-year stints at universities in Canada and then became unemployed. Like most of us, I suspect, I wanted more than practically anything to do mathematics at a university. I accepted the offer of employment here in Barbados without thinking. To be clear though, I do not regret the move at all, but it is not without serious implications and hopefully provokes some thought for the CoYM readership.

I suspect many of us take jobs at a university where the research environment is anything but stimulating. I really do have to work hard here to keep motivated and to keep up with what is going on mathematically in the rest of the world. In fact, just this week we are getting e-mail fully installed in our offices. Up to now, there had been a single terminal from which the entire campus sent and received mail. Barbados is something between a third world country and a developing country. Part of what this means is that academia and mathematics have almost no meaning in the public eye and, to a large extent, no meaning in the eyes of many who work at the campus. Keeping up with my research is one of my most consuming day-to-day concerns here. Of course, we all have this concern. But, I would like to emphasize that there is a substantial difference in attempting to forge on in one's work at any medium-sized or even small university in North America, and having to do the same here in a mathematical vacuum. I would like to think this is a meaningful topic of interest to those of you who, like myself, were willing to do darn near anything to end up at some university doing mathematics.

I'd like to close by emphasizing again that I do not regret my choice of moving and setting up shop here in the sunny south. There are, in fact, many good aspects of living here. Some are actually outstanding: the weather, the pace of life (*i.e.*, the small life outside of the office), distances are short, time stands still, it is easy in some ways to have visitors (if they don't mind the expense). However, not a day goes by when I don't suffer in some sense because of the academic isolation and because of the frustrations which are implicit in living in the Caribbean. It's always interesting and predictable when I'm home in Canada or at a conference and get to chatting with someone about academic life in Barbados. Once they get beyond the sun and beaches and all that and start thinking, as I do, about the day-to-day, then the points I've made above become more clear.

Issues at a One-year Position: The Temporary Blues
by Kevin Charlwood

Here is one man's view of the "one-year position" situation. Keep in mind, these items can vary greatly from institution to institution.

First, the salary issue. I'm being paid $29K, whereas a tenure-track individual had been hired here the previous year at $34K. A whopping big difference, to be sure! $29K is definitely at the low end of the scale. I won't beat around the bush here ... it's just plain low. I should also hasten to add that I was allotted $1K in moving expenses, most of which I used. But even so, those of you who will be looking for a position should bear in mind some of the numbers which may be thrown your way. If you don't get an interview until June or July, you'll be cornered by the powers-that-be. They know they've got you right where they want you in terms of their bottom line, so prepare for that ahead of time. Granted, some temporary instructors are making more after being very late hires, but your chances of getting the salary you deserve are greatly diminished. Late in the search year, you basically have no leverage, unless you're fortunate enough to have another offer sitting on the table. If so, my hat's off to you!

Second, benefits. Bradley is picking up 2/3 the cost of my health insurance, which leaves me to pay around $45 a month. Not bad, as I know some people who are tenure-track and have to foot the entire cost of their health plans, even for HMO's. The downside (ah, and there are plenty!) is that I couldn't get into TIAA-CREF or any other plan offered here. So, I'm on my own completely to save for retirement for now. I figure based on what schools normally give, I'm losing around $2K toward retirement. Ok, I'm still 30 years young (!) but I'm resigned to losing a year in terms of benefits. Keep this all in mind when considering an offer ... you may regret it all later if you don't!

Third, the teaching load. In CoYM last year, there was an abundance of stories relating people's high teaching loads, *i.e.*, more than the standard twelve hours per week. Of those I'm aware of this year, I know of no one suffering through this, but I'm also quite sure it exists. Here at Bradley, I'm teaching 12 hours of Business Calculus (3 sections). I must say that for me, it's an ideal load, as the preparation work is minimal. (That's part of the reason I have enough time to goof off and write some opinion pieces for your reading enjoyment.) I have ample time to dissect my thesis and attempt to work on some related results to make myself more marketable. If you get stuck with four courses and three preparations someplace, that all goes out the window in a hurry, believe me, especially if you haven't had a lot of "full control" teaching experience and experience preparing decent tests. It can really eat up your time if you're not careful, and if you're amongst those unfortunate ones teaching more than 12 hours (like 15, for instance) it's really unavoidable.

Next up: Having to look again. This is perhaps the most professionally crippling item of all. When I consider the sheer number of hours I spent researching schools and putting together my 137 application packages over a period of 8 months, I get physically numb! And now I, like many others of you out there, am staring down the barrel at having to do all this again. But since how you sell yourself to prospective employers is of paramount importance, you'll need to spend oodles of time doing it and doing it well. For some schools, it'll mean the difference between the top 50 and getting into the final list for an on-site interview. Of course, we all want to land that elusive tenure-track job at a good school, and so with 600+ applicants in many cases, it behooves you to spend your time applying. Between now and March, I honestly expect to get little else done besides look and do a great job teaching. If I get two papers submitted out of my thesis, I'll consider it miraculous.

Finally, an intangible: Moving. This can be a pain, indeed! Moving around the country, taking temporary positions (like a gypsy) and not having any chance to put down some roots in a community also takes an emotional toll. The up side is, if you don't like where you are, well, by golly, you'll be out of your current situation in just a few months. The situation is worse if you're dragging a spouse around with you. Don't spouses just love that?! Fortunately for me I'm still single, so I can pick up and take off without any hassles imposed on anyone else. It's still a hassle for me, but I'd hate to be the cause of someone else's discomfort ... especially a spouse's.

Are there any solutions to this "temporary" insanity? I liked the idea that some guidelines be sent to all schools regarding fair and/or reasonable employment practices. [ed. note: The AMS does indeed currently mail a copy of the statement *Supportive Practices and Ethics in the Employment of Young Mathematicians* to each employer who advertises in the EIMS.] Would unionization help or hinder the "temporary" syndrome? Can the AMS or some other such powerful body intercede when the situation gets out of hand? Right now, as the market is so competitive, we as individuals have little voice, so the likelihood is, these temporary situations will only increase, in my estimation. Schools get the best of all worlds: cheap, talented labor that is quite expendable once the contract expires.

References

1. E. Aboufadel, *Job Search Diary*, MAA FOCUS **12** (1992), no. 5, 6; **13** (1993), no. 2, 3.
2. A. Crannell, *Applying for Jobs: Advice from the Front*, Notices of the AMS **39** (1998), 560–3.
3. T. Hull, M. Jones, and D. Thomas, *Interviewing for a Job in Academia*, Notices of the AMS **45** (1998), no. 10, 1353–7.
4. T. Rishel, *The Academic Job Search in Mathematics*, Mathematical Sciences Employment Register, Providence, RI, 1998.
5. M. Parker (ed.), *She Does Math! (Real Life Problems from Women on the Job)*, MAA, Washington, D.C., 1995.
6. A. Sterrett (ed.), *101 Careers in Mathematics*, MAA, Washington, D.C., 1996.
7. A. Crannell and A. Jackson (ed.), *Preparing for Careers in Mathematics*, video, AMS, Providence, RI, 1996.
8. *Mathematical Sciences Career Information*, AMS-MAA-SIAM, `www.ams.org/careers`.
9. *Employment Information in the Mathematical Sciences*, published 4 times yearly, AMS, Providence, RI.
10. *Seeking Employment in the Mathematical Sciences*, Mathematical Sciences Employment Register, Providence, RI, 1994.

CHAPTER 2

Industrial Mathematics

There are several themes that emerge when non-academic mathematicians talk about their work. The first is that there has to be a change in attitude from that of "being smart" to that of "doing things". The most immediate "thing to do" is to find out where the jobs are: moving from academia into the real world requires more initiative than picking up the EIMS listings and seeing which schools advertised this year. This chapter includes advice from several people who made that first step of peeking outside the ivy-covered walls. Karen Singer, then a graduate student at Johns Hopkins University, describes how she found summer internships; these are one way of learning more about a field that interests you. Kevin Madigan offers insight into the financial sector—which he briefly investigated while still in an academic job—and then into the actuarial field.

The second theme that emerges is that industrial and financial jobs, far from being consolation prizes, are actually exciting and fullfilling (not to mention well-paid). As Michael Sand points out in the first article, taking a non-academic job can be "a blessing in disguise." This chapter ends with interviews of two senior mathematicians (Tom Davis and Jim Phillips) who have spent their careers in industry.

We'd like to add a few random pieces of advice of our own. First, smaller firms often don't know about NSF Industrial Postdoctoral positions. So you can show the kind of initiative most companies like to see by gathering information from the NSF and then asking the firm whether they would be interested in having you as an NSF Postdoc. Second, many national labs regularly have internship positions and other jobs for mathematicians—don't forget about Uncle Sam during your job search! Finally, if you are working with your placement office (and if you are interested in non-academic jobs, you should be), contact them in October and November and get on the interview lists for summer employment.

Life After Academia
by Michael Sand

After obtaining my Ph.D. from Berkeley in 1994, I was fortunate to get a one year position at UC Riverside. I got an extension for another year, but that was as

17

long as the university would extend the type of position I had. I went through the motions again and applied to 180 schools. I landed one interview for a job at a tiny school for which I was turned down. I was relieved, because by this time I was not so sure I wanted to stay in academics. The prospect of teaching three or four classes a term, maintaining a research program and earning an academic salary for my efforts was rapidly losing its appeal.

Having decided to work in industry, I managed to get a job in a software lab at Hughes Aircraft Company [*ed. note*: now owned by Raytheon Systems Company]. I had essentially no software experience, but I was fortunate to find managers who were interested in my general problem solving abilities as opposed to any specific knowledge I might have.

How is this job different than my two years at UCR? I am now focused on solving whatever particular problem is in front of me, as opposed to worrying about building a career by writing enough papers and going to the right conferences. The work I do now has an immediate effect and there is no doubt about the motivation for looking at a problem. I no longer have to make a judgment about the value to my career of a problem, since essentially the problems are given to me. This aspect of industry bothers a lot of academicians, but not me. I am fortunate to have managers who want to be told if I am getting bored with my work. So far this hasn't happened. I also find that I collaborate with far more people now than I ever did in academics. It takes the knowledge and experience of several hundred people to produce a modern avionics system.

There are other differences as well. One was the interview. Basically, I just chatted with a few people for a couple of hours and attempted to convince them I could be useful. Hughes then made a decision about me within a few days. This is quite different from the standard academic ordeal in which an interview can be a one or two day event and the decision process can last for months. Another difference is the salary. A look at the salary survey in the December *Notices* will give you an idea of the disparity between academics and industry. Another minor point: I found my job with about 25 resumes over a two month period.

The work itself so far has been sufficiently challenging. I have been assigned to a project for which the basic software has been written. My job involves investigating problems in the functioning of our system. Very often the solution involves some modification of the code. Even so, I do not consider myself to be a programmer since the main part of the work is analyzing problems, as opposed to writing code. Programming is a similar portion of my job now as using TEX was in an academic job.

A significant change is that I no longer need to split my time between the two very different activities of teaching and research. Certainly there are opportunities to teach within an industrial setting, but it is no longer part of my main function. This has two benefits. The most obvious is that I can focus my energy on problem solving. Even mediocre teaching requires a fair amount of effort. The second benefit for me is that I was never able to make my idealistic notion of what university level instruction ought to be mesh with the reality of being judged by student evaluations. (Mine always ranged from high praise to scornful disdain.)

Despite these differences, I have found that many of the skills honed in my academic career apply directly to industrial work: problem solving; understanding complex structures; continuously acquiring new knowledge and expressing technical ideas verbally and in writing.

The bottom line is that I am happier in my current job than I would have been in an academic position. For me, the terrible academic job market was a blessing in disguise.

Two Summer Internships in Industry
by Karen Singer-Cohen

For one summerof graduate school, I wished to work in biostatistics. Most such jobs require you to have studied some statistics, and to know how to use statistical software. I applied to a number of pharmaceutical companies, and also to local groups that do statistics for other researchers. One of these was a department of the research/teaching hospital here that has many ongoing projects with other parts of the hospital, and two were private companies that contract their statistical services out to people doing clinical trials. AmStat News posts information about some of the bigger internships, and I also did a lot of informational calling. I received offers from all three of these types of employers. The pharmaceutical companies were able to offer a higher salary, but the smaller groups would have paid me reasonably well. I accepted a job from Merck & Co., a large drug company that hires many Ph.D. and Masters statisticians to work at Merck Research Laboratories in New Jersey and Pennsylvania. Scientifically, and corporately, this is a very good company to work for. Merck's U.R.L. is www.merck.com.

For the following summer, I decided that I would like to work in a job using discrete mathematics. I applied for internships with:

NIST—National Institute of Standards and Technology (government)
Dept. of Applied and Computational Mathematics
Gaithersburg, Maryland
www.nist.gov

SRC—Supercomputing Research Center (computer, engineering, and mathematics work for the Institute of Defense Analyses)
Bowie, Maryland
www.super.org

Bellcore (telecommunications research)
Piscataway, NJ
www.bellcore.com

AT&T Research (telecommunications research)
Holmdel, NJ
www.research.att.com
(The application is by electronic resume of a specified format.)

The telecommunications field is undergoing much change right now, and some internship programs have been cut back to a smaller size than they used to be, but most currently still exist.

For that summer, I ended up working for Bellcore, which I enjoyed very much. During the application process, I was especially impressed with how well-organized the Supercomputing Research Center seemed to be.

I know that the NSA (National Security Agency) in Maryland also offers internships to mathematicians each summer, for both students and graduates. The application process has to begin well in advance (the preceding fall), since it involves obtaining security clearances. NSA's U.R.L. is www.nsa.gov. The Institute for Defense Analyses in Virginia also has a summer internship program. See www.ida.org for more information.

In each case, I began my search process in December or January. A few jobs started to gain potential in March or April, but it was not until April or May that I was offered the job that I ended up taking. Both times, I was persistent about staying in touch with the potential employers. I believe that this helped me, because in at least one case, judging by the late date of the offer, someone else had probably turned down the offer before the company looked to me. Since the managers knew that I was still interested and available, I was next on their list, and I got the job that I really wanted.

I often started by calling a general number for the company, obtained from directory assistance, and asking the person who answered if his/her company was organizing a summer internship/employment program for students. However, all of the offers that I eventually received were from places where I knew of specific people in the relevant departments and contacted them directly. People from industry whom you meet at conferences can be very helpful in this respect. Even if their company does not have a formal internship program that organizes housing and special seminars for interns, it may have funds to hire a summer researcher.

Company web sites can provide job listings, contact information, and information about specific projects that the company works on. If a project catches your eye, mention your interest in it when you write your cover letter.

I hope that this is helpful! I found internships to be a very stimulating and important part of my graduate education.

Seeking Employment in the Financial Sector
by Kevin Madigan

A few months ago I faced the possibility of having to leave academia, and was actively seeking employment in the financial sector. I had heard (as many of you have) that many of the big trading houses like to hire mathematicians. I also saw ads from recruiters posted to the e-math facility, so I thought I should look into it. Below is some generic information which may prove useful to those of you interested in finding non-academic employment. I am no expert in this area. Many of you may know more about this than I do, but I know many of my friends with Ph.D.'s in pure mathematics knew nothing about this until we started discussing it. Read this information with the following caveat: this is only what I was able to figure out in a few months from reading and talking to people in the financial sector. I went on one interview. Yes, indeed, there is money to be made in this area. *However,* this field is very risky and unpredictable. One must realize that this money requires long hours and accompanies a good deal of uncertainty about the future (will I be fired tomorrow if the market plunges or I cause the firm to lose several million dollars?). One must also realize that the financial world is motivated by greed. It doesn't matter how much you know about cohomology or bounded symmetric domains. All that counts is money. If you can't handle this, stay away. I know of several people who went into this field and were downright disgusted by it. Others will have no problem with it.

The first question I am usually asked is "what do they want mathematicians to do?" Well, a lot of things, but one of the things the financial sector looks to mathematicians for is the ability to understand the esoteric instruments that are being traded these days. In particular, they want what they call "quantitative analysts". Many of the instruments being traded these days are so-called derivative securities, meaning a security whose value depends on something else (e.g., an option to buy 100 shares of stock in company X on December 1 for $100. If it looks like the stock price will be much lower than $100 on December 1 then the option is worthless. If it looks like it will be much higher, then the option may be quite valuable.) Derivative securities can be quite tricky, and valuing them may be very nontrivial (someone offering to sell/buy such a security needs to know what to charge/pay for it, and that's where we can come in). At first you may know nothing about finance, but you do know what a differential equation is. You need to know stochastic differential equations, because many of the models used in valuing financial instruments rely on certain stochastic differential equations, some of which resemble the heat equation (leaving aside arguments as to how good these models are). A good grip on statistics and stochastic differential equations should be enough to get you in the door provided you posses the other required skills.

Many of the large houses are willing and eager to hire a mathematician (or physicist, etc.) to spend time reading and learning the business end while also doing some computer work. For this purpose it is essential that you can program in C or C++ (or both). One of the reasons they are hiring you is that you can serve as an interface between the systems people and the traders. You need to be able to tell the computer to run models, and universally they want you to do this in C and/or C++. It would be a good idea to work through a book such as John Hull's "Options, futures, and other derivative securities", paying attention to running a few models. Many of you already know how to do all of this because of applied mathematics/statistics backgrounds. Those of us in more abstract areas need to learn a little more mathematics and hone the programming skills. One must be prepared to deal with people with no "higher" education running the show and demanding performance in unrealistic time frames. One must be able to deal with the pressure and be able to handle a superior who demands immediate solutions to questions which require several weeks of deep thought if one were to give a mathematically rigorous solution. On the positive side, this is a stimulating and quickly paced working environment, which many people enjoy.

The most important question is "how do I get one of these jobs?", and that involves networking. From what I understand, all these jobs are filled through recruiters or personal contacts. Talk to professors and colleagues and find out who they know in this field. Somehow you have to get a recruiter (or someone on the inside) to want to place you. Remember that recruiters get paid only if they place you, and they may lose interest in you after a few weeks. That is the nature of the business; it has nothing to do with your validity as a human. (Also be wary of recruiters sending you to the personnel office or charging you money.) If you get no interviews through one person after a few weeks, start working with someone else.

The best advice I can give you here is to find someone you can trust who can help you get an interview and guide you through the process. Friends of professors or former classmates are a tremendous help.

Attitude is incredibly important. You can't come off like you know everything, and you can't have the attitude that you only want to do this until the academic job market gets better. These are kisses of death. You also have to make it clear that you are willing and eager to learn.

Of course, there are a lot of other variables to worry about (dress, personal appearance, communication skills, comportment, ...), but those are not really specific to this area. In short, you need enthusiasm, a knowledge of statistics and stochastic differential equations, the ability to program in C/C++, and contacts to get you started.

So You Wanna Be an Actuary?
by Kevin Madigan

Recently, I decided to leave academia and become an actuary. I haven't actually begun my new career yet; I start in June. I want to share what little advice I can give for those wishing to pursue actuarial work, and to share some of my interviewing experiences. Neither topic can be expounded upon at great length.

Let's start with advice.

Start taking the exams. Actuarial employers are impressed with nothing less. My having a Ph.D. was not as important as my having had sat for two of the actuarial exams in February. The first exam covers Calculus and Linear Algebra, and should be no trouble for any of us. The second exam is Probability and Statistics, again not too terribly hard for us. I really cannot overstate the importance of exams. If you want to take more than the first two, it is a good idea to take exams accepted by *both* the Casualty Actuarial Society and the Society of Actuaries. The addresses and phone numbers of these two societies appear at the end of this article. [Ed. note: This article was written before the actuarial exams changed; contact the CAS or the SoA for more information about the current content of the first two exams. The addresses are at the end of this article.]

Unless you have some actuarial experience, you must try to get an entry level position. Do not fret about the fact that you have a Ph.D. or a Masters degree and are being hired to do the work normally given to someone with just a B.S. or B.A. Within a few years it won't matter. The starting salary for someone with two exams passed is comparable to the average starting salary for an Assistant Professor of Mathematics with a brand new Ph.D. In 1993, the average starting salary for a new Fellow (one who has passed all exams and other requirements) was $70,000+.

How do you get an interview? The SoA has a little booklet they will send you with the names and contact information of companies looking to hire entry level folks. Its called "Actuarial Training Programs". Get it. Also, it is imperative that you *network*. Talk to the career counselors at your school. Talk to friends and colleagues. Many of you already know people in the business; they may have taken calculus from you, or have been a colleague in graduate school, *etc.*. Does your graduate or undergraduate institution have an actuarial program? It might not be a bad idea to look at the CML (by geographical area) and do a little cross referencing, looking for members of the MAA, AMS, or SIAM who work in this field. Talk to anyone and everyone about this to see who can help.

I should illustrate this with an example. After making the decision to become an actuary, I found out that one of the guys writing letters for me is the director of the actuarial studies program in his department. We spoke, and he gave me a name, which led to an interview which led to a job (thanks again, Carl!). The only other interview I had was made possible by my father introducing me to an actuary who has a Ph.D. in Mathematics. (My father is not an actuary, but is in the insurance business.) In neither case did I blindly send a resume to someone. They knew it was coming and made sure I was interviewed.

I don't know how useful recruiters are for entry level actuarial positions. I only contacted two recruiters. One flat out told me she couldn't help me due to my inexperience. The other asked me to mail my resume. I haven't heard from him since. I am under the impression that most actuarial recruiters are only interested in placing people with experience. Of course, it won't hurt you to go ahead and contact a few. It won't cost anything (except maybe a long distance phone call), and they will at least know who you are. You may get lucky and find one who knows of a job for which you are well suited.

Another gem of advice I can give you is that actuarial work is not all about mathematics/statistics, it is about business. Do not try to go into this if you expect to sit around doing mathematics all day. Certainly, jobs exist where one does a lot of number crunching and computational statistics. Bottom line, however, is business. You will be sorely disappointed if you expect otherwise. (I am sure there are exceptions to this rule, but as a general rule I think it is a good one. At least, that is what my actuarial friends tell me.) A moderate business background can really help you. Good communications skills are an essential part of many actuarial jobs. If you are a good teacher, pump that up. Some people even claim that there are distinct personality differences between those in the life and casualty sides of the business. It wouldn't hurt you to learn the basics of life, property and liability insurance.

In fact, you really should learn a bit about the insurance industry before trying to get an interview. Of course, you should ask a lot of questions once you get there. You have to impress the interviewers with your seriousness. This is a big career shift. There isn't much advice I can give on interviewing other than the generic. However, I do need to share one important point. Lots of people out there have the idea that a Ph.D. will be unhappy with or unable to handle the mundane aspects of an office job. In fact, this was a big issue during my interviews. I would go so far as to say a few of the people interviewing me had no interest in me at all because of this prejudice. Fortunately, I was able to convince enough people otherwise; I got the offer I really wanted (and I still won't have to wear a tie!).

We all know—or should know—that in order to get a job you have to project a good attitude, convince the people doing the interviewing that you will work well on a team (incredibly important!) and that you will be valuable to the company. It may not be obvious to them that you will be valuable. Do you expect to sit around all day doing the mathematics/statistics that you find interesting? Or do you expect to do your job well? Doing your job well may not entail much mathematics or statistics. However, you posses a very well trained brain, you have to convince those doing the hiring that you *want* to use this brain to help them make money (that is what it is all about, isn't it?).

Here are the addresses and phone numbers:

Society of Actuaries

475 North Martingale Rd

Suite 800

Schaumburg, IL 60173-2226

(708) 706-3500

Casualty Actuarial Society

1100 North Glebe Road

Suite 600

Arlington, VA 22201

(703) 276-3100

An Interview with Tom Davis at SGI
by Tom Davis and Wendy Alexander

This interview occurred in March of 1995, and therefore some of the numbers are out of date—the number of employees at Silicon Graphics, and the performance numbers for what would be considered a high-end graphics system, for example.

Prologue. Just as general interest, here's my background: I've got a Ph.D. in mathematics from Stanford, and did a post-doc for about 3 years in Electrical Engineering, also at Stanford. I taught some mathematics and computer science in small colleges for a couple of years, and since then, I've worked in industry—first for a company that did an office automation system, and for the last 13 years at Silicon Graphics where we're most famous for our graphics workstations. Silicon Graphics employs about 5000 people world-wide, of whom perhaps 1200 are in engineering— about 700 in software and 500 in hardware. I work primarily on graphics software.

Some of the comments below may seem a bit negative, but I really love math, and always have. I still teach some volunteer mathematics classes, read mathematics books, work on problems/puzzles, and so on. It's just that a lot of the things I love don't help anybody make any money, so industry's not particularly interested in paying me to do them.

Remember also that my point of view is pretty limited—my company builds and sells computer hardware and software, so the sorts of things I recommend below are highly biased in that direction. We don't build bridges or design wonder drugs or make rockets.

The Interview.

1. If you were beginning graduate school in mathematics today, what courses would you take to be as valuable as possible in the job market a few years from now?

It would probably depend on how smart I was, and what I wanted to do. Obviously, if I were one of the top two or three mathematicians in the country, I'd be virtually certain of getting an academic job wherever I wanted, so I could stick to pure mathematics if that's what I liked.

For the rest of us, however, it seems clear that even if you want to be in academia, the future job market there is pretty uncertain, so it would be wise to take some more applied courses, or better yet, get some experience via summer jobs or something in industry.

So outside of mathematics, the most useful courses I took or audited were in computer science and electrical engineering—particularly those that required me to build something or make some software work. I actually did sit in on a lot of physics, chemistry, and biology courses as well because they interested me, and it's sometimes useful, when dealing with customers, to know something about what they're trying to do.

In computer science, I'd recommend introductory courses in a few basic areas— it's nice to know how a compiler works, how an operating system works, how to write solid, production code, and (at least at Silicon Graphics) something about graphics and image processing.

In electrical engineering, it's nice to be able to read a digital circuit diagram, although there's certainly no need to know how to design one. Some simple knowledge of the hardware architecture or your computer can make a 10 times performance difference in your code.

For mathematics courses I recommend, see the next answer.

2. *What are the mathematical tools that you use most? What are the things I should be getting familiar with? Is there a particular computer language or skill that is absolutely necessary in a new hire?*

I've worked in a lot of areas of computer science, and until I began working with graphics, I used very little mathematics other than basic principles of logic and deduction. It's a good idea to have some idea how to estimate the performance of an algorithm, which can be tricky.

In graphics, I use linear algebra heavily, projective geometry, and some differential geometry. Other people here who do a lot of image processing work know a bit about the various filtering techniques (Fourier transforms, finite Fourier transforms, wavelets, and so on). I'd guess that in the next few years, there will be a lot more work in image processing because it's finally getting cheap enough to do quite interesting things on relatively inexpensive machines.

Some of the work here requires a bit of numerical analysis, and a lot of our customers use it heavily. We don't do any differential equations work here, but a lot of customers do, so it's good to have some expertise in that area in-house. These folks will deal with the customers and with the hardware designers and make sure we're building the right stuff. Actually, the folks here that do this are not mathematicians—they're often physicists or mechanical engineers who have had huge amounts of practical experience with PDE's.

3. *Suppose you are talking to a brilliant, new mathematics Ph.D. with a "pure" background who really wants to get a job in industry in the next year or so. What minimal re-education path would you suggest?*

If you're a brilliant Ph.D., you presumably are smart enough to teach yourself. I'm not brilliant, and that's what I did. I bought books on the topics I mentioned above and bought a cheap computer, and was my own professor. I wrote a simple interpreter, compiler, and operating system basically from scratch, and it's amazing how much I learned. I bought chips and built some simple circuits, including flip-flops, adders, counters, *et cetera*, which also proved to be extremely valuable.

The main problem with the above scheme is that it's hard to get a job without "official" certification—none of this stuff appears on any of my transcripts—so you should not count on starting in your ideal position. Get any computer job in a company that has jobs like the one you really want, and after a year or two, if you prove to be competent, you can move to any of those jobs. In industry, after a couple of years, nobody cares what your degree is or where you went to school—all that matters is how well you've done there.

My first non-mathematical job was in a marketing group, teaching customers to use a sort of complicated word processor. I was able to get the job because I had some teaching experience as a graduate student and in a junior college. Once I was in the company for a while, it became obvious that I had some technical ability as well, and I never had any trouble getting a job after that.

So my advice is that if you want to change careers, get whatever related job you can, and do a really good job. Also, read as much as you can about software engineering on the side—there is a huge amount to learn, and it's not easy to "derive it from first principles".

4. How many Ph.D.'s from mathematics are in your employ? How many statisticians? Physicists?

I think there are 4 mathematics Ph.D.'s here. I'm a graphics hacker, another of us writes compilers, but the two others actually do some mathematics. One works on problems of surface description and algorithms, and the other works a lot with numerical accuracy concerns for various algorithms.

I actually do a little mathematics, since a lot of people know I've got a Ph.D., it seems like once every couple of weeks or so somebody comes by with some problem that has a mathematical component and I "consult" on it.

5. How many hours per week does the average industrial mathematician work?

Here, at least, we're all just "engineers". When there's no crisis, it can be a standard 40 hour per week job. In a crisis (of which there are plenty), it can be 50, 60, or 70 hours per week. I find that by being clever, I can anticipate a lot of crises and head them off, and I probably work 45–50 hours almost all the time. But when you're new on the job, and trying to impress the boss, it wouldn't hurt to put in some extra hours.

6. How is the work structured? Do they just hand you a project and say, "This is your baby; get it done in 10 weeks?" Or do you always work in a group? How big are these groups? Is there always a deadline, or do some groups work on a more open-ended timetable?

Almost all work is done in groups, but the group is just handed a project and told, "Make a machine that can draw a million triangles per second, costs us less than $10,000 in parts, and have it done in 18 months." This may not be typical of the industry, however.

7. How do you find the mathematicians you hire? Is there a bulletin board where openings are posted?

We never look for mathematicians; in fact, we rarely look for any degrees. We have a job opening like "experienced compiler writer", or "entry level operating systems programmer", and we look for that. Of course for jobs like those, it's more likely that the hiring managers would look at resumes of computer science graduates (at least among people with no work experience). If the applicant has work experience, that is what a manager would examine first.

8. Suppose you are looking to hire an expert in PDE's and, after taking a look at the market and realizing you have a hundred people to choose from, you start thinking about what other attributes you want this person to have. What's on that list?

I'd make sure the person we hired had experience with solving them on big machines, and had a good idea of how to deal with unreliable hardware. For giant PDE's, we usually use networks of multiprocessors, and stuff is breaking all the time during the calculations. The algorithms must be robust enough to survive this. Of course, I'm only familiar with the sorts of work we do, but it typically involves huge problems requiring super computers.

9. If you could form your own applied science department to train future employees, what would you put on the curriculum?

Actually, for our needs, computer science and electrical engineering departments do a pretty good job at supplying 99% of our engineers.

An E-mail Interview with Jim Phillips at Boeing
by Jim Phillips and Wendy Alexander

1. If you were beginning graduate school in mathematics today, what courses would you take to be as valuable as possible in the job market a few years from now?

I find this difficult to answer without talking about a complete curriculum, and I'm not informed enough to lay such a thing out. A Ph.D. going into industry needs breadth (a good foundation that gives him/her the capacity to dig into a variety of areas of mathematics and to "think mathematically"), and depth in *some* area. In general, most graduate programs provide both, between preparing students for qualifying or preliminary exams, and the Ph.D. work itself. Beyond that, one needs to develop an understanding of where applied problems come from, and how one can get a grip on them and solve them. If you knew you were going into a particular industry, it would be easier to say what to take to accomplish this. Lacking that foresight, some solid courses in something like engineering physics and mathematical modeling would be desirable. To learn more about common tools to use, a good numerical analysis foundation is a must. Slightly (but not much) behind this, I would list optimization and statistics. Since most physically based problems start with a PDE (or at least an ODE), some good background in those is most desirable. Finally, some background in scientific computing issues (*e.g.*, basics of data structures, solving large scale problems on computers) is highly desirable. I realize that even this basic list is a long one. One implication of it is that it is hard to take courses in a large number of these things if your primary interest area does not have a large overlap with them.

2. What are the mathematical tools that you use most? What are the things I should be getting familiar with? Is there a particular computer language or skill that is absolutely necessary in a new hire?

In the environment my group is in at Boeing, numerical analysis is probably the tool bag used most. I listed desirable "things to get familiar with" above. There is no particular computer language or skill that is an "absolute" necessity. The overall requirement, however, is that you have solid experience using modern computing tools to solve non-trivial problems. The most widely used tools vary with time. Unix is not as important as it was a few years ago, while the use of C++ and Java is growing. Fortran and "C" are the languages used most in our work. Experience with math software libraries and problem solving tools such as Matlab and Mathematica are also helpful. Someone with a reasonable computing background can generally pick up other computing skills as needed.

3. Suppose you are talking to a brilliant, new mathematics Ph.D. with a "pure" background who really wants to get a job in industry in the next year or so. What minimal re-education path would you suggest?

Find a way to get involved in some real world problem solving, then work your tail off to fill in the background you need to solve the problems at hand. A postdoctoral position at a national lab or industry would be one such possibility.

Another possibility if you are at or near a university: start spending time talking to colleagues in engineering or the sciences, and get involved (by volunteering, if necessary) in solving the mathematics-related problems that come up in their research. While you are in limbo before finding such a position, work your way through some good books or journals that focus on engineering or physics applications, then develop computer tools (mathematical models, algorithms, computer programs) to re-solve some of the problems discussed therein. Of course, if you are at a university, you can also sit in on courses and seminars that are helpful, too. But I think the key thing is: Get involved in problem solving. Industrial mathematics is neither a spectator sport nor one that you can experience by reading books or understanding more theoretical mathematics.

4. How many Ph.D.'s from mathematics are in your employ? How many statisticians? Physicists?

Ph.D.'s: Math, 39; Statistics, 12; Physics, 2; Engineering, 7; M.S., 34 total.

5. How many hours per week does the average industrial mathematician work?

40 is the standard expectation. Many work more, either because that is their professional style, or perhaps because they are carrying their work further on their own so they can get a paper or conference presentation from it.

6. How is the work structured? Do they just hand you a project and say, "This is your baby; get it done in 10 weeks?" Or do you always work in a group? How big are these groups? Is there always a deadline, or do some groups work on a more open-ended timetable?

There is a great deal of variety here. I think that it is accurate to say that most mathematicians in industry are part of groups dominated by folks with engineering or science backgrounds. As such, the mathematician is part of a team working on a project of interest to their employer. The "mathematical part" is generally thoroughly entwined with the other issues; it cannot be easily separated. In particular, it is rare that the mathematician will be handed a problem, sent off to solve it, and asked to bring back a solution on a platter. (When that does happen, the experienced mathematician will immediately start asking questions, because he/she will immediately suspect that the problem they were asked to solve is not the REAL problem that needs solving.) A team working on any particular problem may vary from one to perhaps six or eight. There is often a hard deadline; it depends on whether the problem under consideration is part of a schedule driven project (*e.g.*, getting a new product to market by a target date), or whether it is part of longer term research and development. In the environment I am in, the mathematicians often work as consultants around the company. Their customers may contract to have the mathematician support them or their project at, say, a half-time level for a given period of weeks or months. Whether the support is continued beyond that depends on the state of the project and the need for further mathematical help with it.

7. How do you find the mathematicians you hire? Is there a bulletin board where openings are posted?

Our usual sources of advertising positions are ads in SIAM News, and announcements to e-mail distribution lists such as NA Digest (`www.netlib.org/na-net/na_home.html`), a network of people interested in numerical analysis issues.

8. Suppose you are looking to hire an expert in PDEs and, after taking a look at the market and realizing you have a hundred people to choose from, you start thinking about what other attributes you want this person to have. What's on that list?

Experience in, and a real interest in, solving real world problems. Ability to interact with people and work in teams. Good communication skills, both verbal and writing. Mathematical breadth, with depth in the PDE area (since that is what you hypothesize that I am looking for). The depth would include knowledge and experience in solving nontrivial PDE problems numerically, and a good understanding of some class of physical problems (*e.g.*, fluid dynamics or electromagnetics) that gives rise to PDEs.

9. If you could form your own applied science department to train future employees, what would you put on the curriculum?

This gets back to question 1, or at least my answer to it. That is really a tough question, because it is so tempting to list every theoretical and applied area that one might use in his/her career. But in actuality, the details of what one takes formally are not as important as gaining the overall breadth, and the depth in one area, as I mentioned before. After that, the interest in, and experience with, solving real world problems, and the attitude toward real world problem solving that one brings to the job, outweigh specific curriculum requirements. Finally, if I were forming the department, the faculty hired would all recognize that their graduates can have interesting and fruitful careers in nonacademic settings, and that such careers are in no way second class or less desirable than academic careers. They are simply a second career option.

References

1. *Careers in Science and Engineering*, National Academy Press, Washington, D.C., 1996.
2. *Mathematical Scientists at Work*, MAA, Washington, D.C., 1993.
3. *Mathematical Sciences Career Information*, AMS-MAA-SIAM, www.ams.org/careers.

CHAPTER 3

Life in Small Schools

Much of the readership of the CoYM were academic mathematicians who were at—or who hoped to wind up at—an institution smaller than the one that granted their Ph.D. Although it was also true that a large minority of the readers were at major research universities, no one had questions about how to obtain a job at such a place, or even how to succeed once there.

On the other hand, the transition to a smaller university or college is fraught with questions and misunderstandings. In this chapter, we present a small sampling of the variety of academic institutions where mathematicians work. We do so in descending order of the level of degree offered, from University to Four-Year College to Two-Year College. Several themes emerge from these examples—the need to document teaching ability, the importance of service to the institution, and the necessity of understanding the mission of the institution. These themes will reappear in the chapter on tenure.

We end with a small essay by Evelyn Hart on one response to the frequent remark that professors have it easy with an n-hour work week (if they teach an n-hour course load).

Mathematics at Smaller Universities
(or "Is there life after the Ph.D.?")
by Paul Shick

Given the condition of the job market, many new doctorates who have spent their entire academic careers at predominantly research-oriented institutions will be getting jobs at smaller universities. Having talked with a number of graduate students facing this seemingly mysterious fate, I thought it might be useful to give a quick summary of what life is really like at a typical small university (or at least my own experiences at such places.) As a veteran of 10 years of university employment spent mostly at smaller places, I might be able to give a fairly realistic view of what really goes on. This is not, however, intended to describe *all* smaller universities, and I apologize in advance to those who feel that I have maligned their institutions.

To put this in perspective, I am writing this at John Carroll University, a Jesuit institution located in the near eastern suburbs of Cleveland. We are a fairly selective school that emphasizes undergraduate education above all else. We have roughly 3200 full-time undergraduates, with about 4500 total students if you add in part-timers and graduate students from our various masters programs. As a Jesuit school, we put a lot of stress on a strong liberal arts core of courses, with a lot of contact between faculty and students.

What's the basic job description?. Teaching, Research, and Service. The official description says this:

1. Teaching: A new tenure track faculty member will generally teach an 18 hour load, spread over 2 semesters. (Most of our older faculty who are not active in research teach 24 hours.) Since some of our Calculus courses are 4 hours and our M.S. courses count as 4 hours (even though they meet for 3 hours a week), this usually works out to 3 courses in the Fall and two in the Spring. Typically, a lower level course like Calculus or Elementary Statistics will have about 30 students and the instructor will have help, either from a TA or from an undergraduate homework grader. Upper level undergraduate courses vary from 8 to 30 in enrollment, depending on the popularity of the instructor and whether the course is a requirement for math majors or an elective. M.S. courses carry between 5 and 8 students, usually. A second part of our "teaching load" is academic advising. Typically, faculty members advise about 10 Freshmen or Sophomores (from many different majors) and a few undergraduate math majors.

2. Research: To get tenure at a place like this, "the candidate must show evidence of continuing scholarship." We have no formula for publications, etc., that suffices, although other departments on campus do. What's required is that the candidate remain active and demonstrate it, preferably by the usual route of publishing in refereed journals, but other means (lectures at seminars, meetings, etc.) can work, too.

3. Service: Typically, faculty at smaller places get more involved with the running of the university than those at larger schools. One is expected to chip in and help, but not necessarily be a "political animal."

What's the reality of the job?.

1. Teaching/advising: At smaller schools, teaching is taken quite seriously, and one is expected to do it well. All our tenured faculty, for example, will sit in on classes of the untenured members of the department at least once a year, to evaluate and offer advice. Preparation and classroom teaching are essentially the same as at larger institutions. What really happens, though, that grad students often don't expect, is that one spends a LOT of time with students outside of class. Students at smaller schools generally expect and get more personal attention from instructors than those at larger places. This places greater demands on the faculty, but there are rewards to it as well. The amount of time spent in advising illustrates this point, also.

2. Research: The real question here, I guess, is whether it is possible to do serious mainstream research at a typical undergraduate university. The answer is a provisional "yes". One can indeed remain active mathematically, but it takes a great deal more effort than at larger institutions. First, the odds are that there won't be another person with whom you can collaborate in your own department, and you'll have to work harder to find someone to bounce ideas off. E-mail helps immeasurably in this regard. Traveling to meetings or just to spend time in other departments is a must. Second, there may be financial compromises required, too, in trying to be an active researcher. I absolutely avoid teaching during the summers so that I can get some real mathematics done then, despite the fact that summer school teaching would be more lucrative. Third (and perhaps the most obvious) is

that there's less time for research than at larger places. It's quite possible to spend an entire week working hard and realize at the end of it that you didn't have any real time to do mathematics. One has to be conscious of this and act accordingly, which isn't always as easy as it sounds. Being a popular and accessible teacher can be seductive, making it easy to forget just how much time you're spending with your students. On the other hand, hiding yourself away for long stretches to do research seems contrary to what a smaller university is all about. Striking a balance between these (occasionally conflicting) goals is hard to do and will certainly involve feelings of guilt. Finally, there's a question of what sort of research one can do at a small place, too. In physics, for example, one can do experimental holography or something along those lines, but experimental particle physics is clearly impossible. In mathematics, many people seem to resort to "nook and cranny" research, looking at smaller problems that are new but not really mainstream.

I'm somewhat convinced that in order to do truly mainstream research in a field, over a long period at a smaller place, one must work in collaboration with others who are in larger places. This may involve being a "junior partner" with some very bright people in certain projects, but I find this to be a rather stimulating (if often humbling) role.

I find that I've emphasized the negative aspects of research so far. There are a lot of plusses, at least in my limited experience. First, my administrators seem to be quite aware of our relatively isolated position and have made it easy for me to get travel money. Second, we have access to locally funded "Summer Research Fellowships" that can help ease the monetary distress alluded to earlier. Third, in some ways, there's more appreciation shown for those active in research at small places than at large institutions. This isn't true, of course, for all such places, but I've found a lot of support and encouragement from both the administration and my colleagues in the math department. In particular, some of our senior people who have absolutely nothing to gain by it have been very generous in covering classes so that I can go to meetings, *etc.* Given that our university's fairly recent emphasis on research tends to forget the senior people when it comes to promotion and salary, this is a very pleasant development. I must add, however, that some of my friends at other smaller places have not shared my happy experiences in this matter. Finally, the lower expectations for publishing at a smaller place can be quite liberating. Rather than looking for problems that I'm fairly certain I can make progress on (as some friends in higher pressure positions have told me they do), I can work on literally anything I want. This has allowed me to work toward some very neat conjectures (one of which may actually be worked out soon, I hope) that I probably would have shied away from had I been laboring under the "publish or perish" pressure that you've all heard about.

3. Service: Unfortunately, I'm one of those people who gets annoyed by things and, rather than ignoring them, tries to change them. This has gotten me involved in far too many university committees and projects. The positive side is that one can have a real effect on the way the university is run. The negative side, of course, is that this is just one more way to fritter away time that might be spent more productively. Again, some balance must struck.

One remark that I must make, in all honesty, is that the descriptions above apply only to tenure track positions. Visitors are typically treated much more shabbily, with higher teaching loads and less support for research and travel. On the other hand, the university does not generally expect research, advising, or service from visitors, so that there may not be all that much extra work for a visitor, even with the higher teaching load. Still, our administration regards visitors as short-term "calculus slaves," despite the efforts of our department to convince them otherwise. Such a visiting position is not the ideal way to begin an academic career.

Conclusion. It is possible to remain an active mathematician at a smaller university, although there are more demands on one's time than at research-oriented places. Further, there are a lot of rewards for the extra work, as evidenced by the fact that many people actively chose such positions back in the old days when one could make choices about jobs. Finally, for those who feel forced into such a position by the job market, it is possible to move from a smaller university to a larger, more research-oriented place, if one remains active. I have a few friends who have done this successfully, although the number is small. I'm honestly not sure if the small number is due to the difficulty of such a move or if it's because most people—like me—are genuinely happy with their positions at smaller places.

Liberal Arts Colleges:
What to Expect and What is Expected
by Stan Wagon

I offer here some thoughts to young mathematicians seeking tenure track positions at liberal arts colleges. I have been on many search committees over the years and seen very many candidates.

Expertise. In a department of about ten people it is helpful when individuals can take responsibility for curricular and other issues pertaining to their specialty. It is often said that a competent professor ought to be able to teach any part of the undergraduate curriculum. But if "teaching" means "teaching with a personal and modern point of view", then this is simply false. Thus a certain natural division of labor takes place. Thus I would expect a candidate to be willing and able to discuss curricular issues related to advanced and elementary courses in his or her field, to have a knowledge of the important textbooks, to be aware of innovative approaches to the field using computers, and so on. In short, most departments will be willing to defer to the specialist, but they must have confidence in him or her.

Workload. A mathematician at a liberal arts college is very fortunate. The atmosphere is lively, the students are diverse and talented, and one gets to know colleagues in other departments very well. A faculty member gets to teach a variety of courses, and the liberal arts ethic infuses the curriculum and the teaching methods with a certain spirit that is liberating. But this means that there is a lot of work to be done, and new doctorates are often woefully unaware of how heavy the workload can be. The top colleges all demand that professors do research as well as teach several courses each semester; and the students at these colleges can be very demanding—usually properly so.

When some of these courses are new to the instructor, the workload can be overwhelming. New instructors will usually get high priority in choice of courses. In the first semester, they should not take on courses involving material with which they are not familiar. After a year or two, they can take on such courses more often, to learn well as much of the curriculum as is reasonable.

Some colleges put young faculty on important and time-consuming committees. I feel that any junior faculty member should be free to say "No" to such requests, reserving his or her service component to the departmental level (advising, curriculum work, and other such tasks). Some college service is a reasonable demand, but serving on major elected committees with weekly meetings is not. Such committee work can be satisfying and productive, but it has a cost. Wait at least 3 or 4 years before agreeing to serve, pointing out that high quality scholarly work cannot be maintained if 40 hours of every week are occupied (and they will be if you serve on, say, a college wide curriculum committee at the same time as you teach three courses). The tenured faculty should staff these major committees.

Teaching. It would be presumptuous to expect everyone to teach in the same style. Each individual must find the approach that works, and just about any style has the potential for success. Steven Krantz's booklet, "How to Teach Mathematics," contains many insights that will be valuable to a fledgling professor. Here are some things that seem especially appropriate to expect at a liberal arts college:

1. Respect for the students and their questions and difficulties.
2. Enthusiasm about the material, together with knowledge of how it is used and why it is important.
3. An understanding of the effect that modern computers have on the theory and practice of mathematics.

And, most importantly, math professors at liberal arts colleges should try to instill a respect for, and confidence in, the mathematical way of thinking. This will be of much more importance to almost all our math students than any specific theorems or techniques they learn.

Scholarly Work. I take a broad view of research, preferring the phrase "scholarly work", which includes expository articles, articles about teaching, books, problem solving, and the like. During the interview process I look for signs that the candidate will have a lifelong commitment to scholarship. When too much time in an interview is spent discussing questions such as: *What exactly is required for tenure? Does this count? Does that count? Are n papers sufficient?*, a red flag is raised in my consciousness. The tenure system has many drawbacks. Nevertheless, we must predict, when hiring and promoting, whether a candidate will maintain a lifelong scholarly career. Thus I look for deep heart-felt feelings about mathematics; I want to feel that the person thinks about math because he or she loves it. That makes it more likely that the person will carry on for 40 years, even if the research direction changes dramatically, as it no doubt will, and also indicates that the person will bring enthusiasm and innovation to the classroom. Despite a general raising of standards regarding scholarship, many candidates feel that a liberal arts college, unlike a university, is low key when it comes to research and, once tenure is achieved, there is little in the way of scholarly requirements. Nothing could be farther from the truth at Macalester and its relatives.

Also, I look for indications that the person will be able to bring his or her scholarly work to the students via honors projects, independent study projects, or advanced courses. While this may not be easy for a new doctorate, I believe it can and should be developed and I look for signs that it will be. If someone arrives with a very advanced research program that has little or no connection to the undergraduate curriculum, then during the first six years there ought to be a subtle shift in the research program so that it is not exclusively high level. Otherwise, the job becomes too much like two separate jobs. I don't mean to say that a strong research program should stop or radically change, but there must be some effort to connect one's scholarly work with the interests and abilities of the undergraduates.

A Final Note. Naturally, in the current job market, it is tempting to take the first job offer that promises the chance of long-term employment. Yet most Americans change jobs regularly, and a new Ph.D. recipient should not necessarily expect his or her first job to last for 30 or 40 years. Think strategically about your whole career, keeping in mind that you will look much better to prospective employers after you have two or three years' experience. Thus, when weighing temporary positions versus tenure track positions, take the long-term view and tilt the scale in favor of the position that offers the best chance of giving you valuable experience and pushing your career in a direction in which you want to go.

Working at Community Colleges
by Leonard VanWyk

I had a tenure track job at a small community college in upstate New York for two years before I went back to graduate school.

The standard load of contact hours was 15. The most advanced class was "Calculus 4." The pace of almost all classes was slow, and the amount of material covered was noticeably less than that of a comparable class at a decent 4-year college. Most of the students were somewhat dim, with a few excellent exceptions. The basic duties were teaching, "service," "community involvement," and "professional development." Obviously, by far and away the most important of these was teaching, which was evaluated both by student evaluations and by the Chair. "Service" involved advising, participating on committees, etc. The "community involvement" component, which increased in importance with the number of years you had been there, involved stuff like giving talks in local schools. "Professional development" had a very loose interpretation.

When I interviewed, I was flown in from CA, all expenses paid. The interview was sometime in April. The year I left, they instituted a rule on reimbursement of interviewees' expenses, which stated that full compensation for travel expenses was not given if you were offered the job and you turned it down. I *think* that NO reimbursement was given in this case, but I am not sure. This was in the late 80's, and it was *very* rare for Ph.D.'s to apply then. I have no idea if this is still the policy there.

(More) Observations on Community College Teaching
by Tim McNicholl

I worked part time for Northern Virginia Community College this summer teaching linear algebra. Here are a few of my observations.

Firstly, the clientele are quite different from a university's. They can be any one of the following: (a) People going back to school for their masters or bachelors degree, who need to take some courses before starting. For instance, I had a large number of people who were going to start graduate school in economics in the Fall. (b) People who are enrolled in a degree program and who are taking summer courses. (c) People taking courses in order to get teacher certification. (d) Recent immigrants who are taking courses in their major until they can get into a degree-granting program

In other words, your teaching will have to satisfy a large number of needs.

Secondly, the courses taught at a community college vary widely. At NVCC, they teach everything from math for auto mechanics to a two-semester course in differential equations. The lower level courses are taught by *all* the staff.

Thirdly, as I mentioned above, there are a large number of recent immigrants who, for whatever reason, haven't been able to get into a degree program at a degree-granting institution, and are taking courses in their major until they can get into a program. A *lot* of them are Asian students who are still learning English. You have to keep this in mind when grading writing assignments and cut people a little slack. Again, an instructor at NVCC works with a very diverse group of students.

A fourth observation comes from a job search the department conducted while I was there. It was for someone to teach primarily remedial math courses. Things that mattered a lot: (a) previous teaching experience at a community college, even part time; (b) student evaluations; (c) experience with technology in the classroom; and (d) evidence of a dedication to teaching mathematics at all levels.

I don't know how they handled the interviewing process.

My fifth observation is that the purpose of a community college is to "serve the community". They were started to help people who, for whatever reason, needed to beef up their academic background before entering a degree program elsewhere. They now serve a wide variety of other needs as my earlier observations illustrate. As many 4-year institutions are now cutting back on their remedial programs, you can expect community colleges to pick up the slack. Thus, I think that the ability to teach remedial math courses, and things such as math for auto mechanics, will be *very* important in the selection process.

Finally, let me share my point of view on the (lack of) reimbursement for travel expenses during the interviewing process. Remember that at community colleges nothing matters more than ability in the classroom. Community colleges have to make sure they have somebody who can deal with the wide variety of students they get and who can handle all levels of mathematical ability. On the other hand, these schools don't have a lot of money (with the exception of Nassau Community College in NJ, the highest paying college in the country!). This is why applicants sometimes encounter the expensive and annoying interview process some of these places use. Because of this, unfortunately, community college jobs will probably go to those who are local. Rather than inveigh against community colleges that use such practices, it would be better to find alternatives to their interviewing process and try to tactfully suggest them (don't appear to be telling them how to run their school). For instance, maybe they would accept a video tape of someone teaching a class as a substitute for the first interview.

Time Spent Teaching Is Only
Part of Time Spent Working
by Evelyn Hart

I am still trying to educate people who tell me that it must be nice having the summers off. When I have the energy, I say that I work in the summers. I tell them that I go from 50–60 hours per week during the semester—that gets them!—to 30–40 hours a week in the summer. I tell them that even though I'm not paid to, I have to work in the summer doing all the things that I'm expected to do but don't have time for during the school year. I do this with a smile, also saying that it's great to have ultimate flex time in the summer. I think I do make an impression by saying these things.

For those who really want to know what research in mathematics is like, it's good to have a copy handy of the wonderful booklet by Barry Cipra *What's Happening in the Mathematical Sciences*. It's from the AMS, has a beautiful cover (which makes a difference!) and describes projects in pure and applied mathematics. It has pictures of the people involved, and is at least mostly accessible to the educated layperson.

References

1. B. Cipra, *What's Happening in the Mathematical Sciences*, AMS, Providence, RI, 1993–8.
2. B. Case (ed.), *You're the Professor, What Next?*, MAA, Washington, D.C., 1994.

CHAPTER 4

Doing Research

Continuing mathematical research away from an advisor was a hot topic for the Young Mathematician's Network. This chapter contains advice on that topic from a whole host of mathematicians. We begin with one of the few pro-one-year-position articles that appeared in the e-pages of the CoYM. We follow this with several of our favorite articles by recent doctorates; these address the questions of how to start a research program and how to keep it going once started (despite committees, students, isolation, and other contrary factors). The fifth article promotes the idea of a mentor (we recommend a research mentor as one of a cadre, along with a teaching mentor, a campus politics mentor, a departmental politics mentor, ...).

The last article in this chapter is our "advice from the pro's" section. It summarizes a Project NExT panel discussion which took place at the January 1995 Joint Meetings; the panelists describe how important it is to stay active in research and how difficult a task it is to do so. The panel consisted of a variety of mathematicians from a variety of schools. The common denominator was that "every one of these folks takes their teaching and their professional work very seriously. (They also take themselves not seriously at all!)"

The Good Side of One-year Positions
by Daniel Lieman

I would like to comment here on the value of one year postdoctoral positions which are intended to be research oriented—or at least, with balanced workloads. The "worst" one-year position I know of was offered to a friend of mine this year: four courses per semester teaching load, at one of the most expensive colleges in the country, paying the grand total of $10,000.

But I think one year research positions can be very valuable in building a career. When I graduated, I was offered the MSRI position, and also spoke with several other schools about tenure-track jobs—none of these schools were willing to defer a job for me to go to MSRI for the year, and so I went to MSRI with nothing lined up after that. There were twenty to thirty postdoctoral fellows at MSRI this past year of 1992–3 (including people who came through for only a semester); most had no support lined up the following year. This is definitely a change from years past.

The MSRI postdocs did quite well in the job market. Many were offered several good jobs; people ended up at places like Chicago, Stanford, Harvard. But I think this was due to the fact that many of us greatly exploited the extra time we had between our first search and our second one. I wrote two papers that summer after graduate school and in the early fall—both were accepted before I reapplied. I also got recommendations from some big names—I think these two factors made a tremendous amount of difference. Of course, one could still write papers in the summer before taking a heavy teaching job, but it really did take a lot of my effort through late September to get the papers out; I couldn't have done that if I had a heavy teaching load.

The other strong advantage of a research postdoc is that you then hit that first teaching job in midresearch. Instead of coming in with a finished thesis, and wanting to get some time to write it up, you come in with your thesis finished, broken into papers and published, and new research begun. That's a heck of an advantage.

MSRI is concerned that people getting the postdocs don't just prolong their agony for a year. There was concern at the highest levels that the MSRI postdocs really did help people start careers. To the extent that the MSRI directorate could facilitate that, they did.

The point is this: some one year jobs are good. Very good. And it's important to continue to distinguish exploitation (whatever the duration) from opportunities, however risky or short term.

Finding a Healthy Research Career at a Teaching Institution
by John D. Lorch

At risk. Several years ago, a close friend of mine quit smoking cold turkey. When pressed for an explanation, he said simply that he felt his life was being threatened, and that drastic measures were needed to correct the situation. I've had several occasions to recall this conversation, most recently with regard to my life as a mathematical researcher at a small, student-oriented institution with a large teaching load. Don't get me wrong. I love mathematics teaching, and I believe that it is extremely important. However, anyone who has a job like mine knows that the day-to-day responsibilities are nearly overwhelming, and the long, uninterrupted hours that most of us need to do mathematics research are hard to come by. Before you know it, your life as a mathematical researcher is at risk, and corrective measures (hopefully less painful than cold turkey) are needed.

Time: More precious than gold. You have the desire to do research. Now you have to find time. You need time to read, to reflect, to do examples, to conjecture, to go to the library, to attend colloquia, to email colleagues, to write, and to be stumped (my personal favorite). I can assure you that none of this stuff happens in a thirty-minute time period. What you really need are large blocks of time, consisting of two hours or more, and lots of them. So, how are you going to find this time, when the other two members of the trinity, Teaching and Service, seem to gobble up 94% of your day? The first order of business is to cut the fat. Here are some pitfalls to avoid (or at least try to take in moderation).

Hall chat. It's almost irresistible. You finish teaching at 2:00 pm and head down the hall where several other juvenile delinquent assistant professors are waiting for you. Someone tells a good story, maybe about a cosine curve that looks like an octopus (see [B] below), and before you know it, the clock has advanced to 3:30 pm. Oh, well. It's too late to do any research now, because you have to pick your daughter up at 4:15 pm. I'm not implying that you should be an island. In fact, all kinds of important information can be picked up through hall chat, but be careful to protect your potential two-hour blocks of time.

E-mail chat. This high-tech version of hall-chat eats at least twice as much time as the original. I have two email accounts: One at home and one at work. I only check my work account three times a week, and have never sent a lengthy message from this account. For the most part, my work account really is for work, whereas most of my longer ramblings originate from my home account.

The teaching schedule from you-know-where. Here's an example: MWF classes at 8:00 am, 10:00 am, and 1:00 pm, with a TTh class at 10:30 am. If you count lunch at 12:00 pm, the only chance for a two hour time slot on MWF is at 2:00 pm. But this is when you are waylaid by your talkative buddies, and so nothing in the way of research gets done on MWF. That leaves TTh. You get up late and arrive in the office at 10:00 am. By the time you are done teaching, lunch time has rolled around. At 1:00 pm you arrive back in the office, ready to do research. However, you have grading from MWF to do, and several students stop by. By now it's 4:30 pm, and time to pick up your daughter.

On the other hand, here's my idea of a perfect schedule: MWF classes at 8:00 am, 9:00 am, and 10:00 am, with TTh class at 8:00 am. This schedule gets me out of bed, and allows for large blocks of time for assisting students, preparing for classes, and for doing research. When the department is making teaching assignments, it is important to be thinking not only of the classes you want to teach, but also about the effect your schedule is going to have on your research.

Summer teaching. Two years ago at a regional MAA meeting, a colleague told me that he was about to embark on a new research project in a field outside of his specialty. His plan, naturally, was to spend a lot of time over the summer getting started in the field, and go from there in the fall. I saw him again at a winter meeting, and asked after his project. He reported gloomily that he had taught some courses over the summer for extra money, and that his teaching had interfered with the new project. Beware of summer teaching. You must weigh the benefits of a (usually meager) summer teaching income against the freedom to develop research/teaching projects at an accelerated pace.

Over commitment. It's so unavoidable that I hesitate to comment on it. Being a first-year professor with a new Ph.D. is a heady experience, and the tendency is to dive into the trinity head first without checking for water. Remember this, though: although your department wants, and rightfully expects, you to work hard for them, they also want your tasks to be well done. Before agreeing to do any optional task, a good rule of thumb is to seek advice on the matter from a senior colleague, and be reasonably sure that you will have the time to do the job well.

Inefficiency. I need note pads, so I go to the departmental office to raid the office supplies. I find that someone has beaten me to it, and no note pads are to be found. So, I head across campus to the bookstore. Forty-five minutes and several conversations later, I'm back in my office, and I suddenly realize that the head of the honors program wanted a hard copy of my new course proposal today. There's no time for campus mail, and the honors guy isn't known for his patience, so I gather the proposal and take it across campus. Forty-five minutes and some serious brown-nosing later, I'm back in my office, and I recall that a faculty meeting will occur across campus in thirty minutes. Defeated, I have just enough time to rest my legs before leaving again. It seems sophomoric, but some basic time-management skills are a must. Every Sunday afternoon, I take about an hour to review my semester goals, and to figure out what needs to be done in the coming week. I then write out a tentative schedule for each day, being sure to block out time for research. This way, the three trips I took across campus become one, and the previously wasted time can be spent on more interesting stuff.

While I'm on my soap box ... , I'd like to mention some other little tidbits that may help a lagging research program.

Research libraries. If you're looking for a job, be sure to find out whether the institution to which you're applying either has or is near a good mathematics library. It's just too easy to take your library for granted in graduate school, and later find yourself out in the middle of nowhere with only inter-library loan to rely on.

Contacts in the field. Chances are, if you're in a small department like mine, no one else on your faculty works in the same research field that you do. This isolation is unhealthy, because it's easy to lose touch with current advances in your field, and because there's no one around to trade ideas with. As a result, maintaining contacts in your field at other institutions is quite important. This can start with your thesis advisor. Ask him for a list of established researchers who are working in areas similar to yours, and then mail them your dissertation and whatever other papers you've written along with a brief note asking (1) for comments on your work and (2) to be put on their preprint mailing list. Sure, a lot (maybe almost all) of these folks won't contact you, but a few will write back and express interest in your work. These are the people that you will want to maintain communication with. Another way to make and maintain contacts in your field is to go to meetings, even if you are not giving a talk. Be sure to come armed with questions for your senior colleagues.

Companion projects. Along with my normal mathematical research, I find it useful to have a couple of other projects, which I call companion projects, in progress simultaneously. These projects are meant to be fun, useful, and less mentally taxing than frontier mathematics. I often use the projects as a vehicle for exposition of some kind, most often in the form of mathematics teaching modules. The document presently before your eyes is a companion project.

These projects can benefit research in several ways. First, they help in avoiding burn-out, since even the most interesting mathematics can become tiresome without another outlet. Also, companion projects are great for those times when you're completely stumped. Not only are you being efficient by making progress on another project, you're also allowing your subconscious mind to work on the obstacle while your conscious thought is directed elsewhere (see [P]). Anyway, after a while you come back to the obstacle with a renewed mind and perhaps a fresh approach.

Put it in writing. I find that the act of writing and rewriting my research and teaching ideas in TEX necessarily clarifies my thoughts and helps me to focus on the main issues at hand. When I'm really stumped on a research problem, being able to share such a document (or a subset thereof) with a colleague helps him understand quickly what my difficulty is, and thus increases the likelihood that he'll be willing and able to assist me. Also, it's good to keep in mind that an idea in TEX is closer to being an article than an idea that isn't.

Guilt. A student just came by my office, and I was happy to admit her even though it's not my office hour. As I mentioned before, I work at a small, student-oriented institution, and such behavior on the part of an instructor is not just optional, it's expected. This leads us to a question that I've heard many times from colleagues at similar institutions: How can we, in good conscience, devote large blocks of time to mathematics research when we could be doing yet more to improve our teaching and service to students? I've heard a variety of pat answers, including the common one about how research makes one sharper and more likely to be an innovative teacher through immersion in the field, as well as the one about how very high level mathematical ideas can sometimes be brought into the undergraduate classroom. I think these are relatively good answers, but they fail to strike at the heart of the matter for me. The happiness that I find in academia is a conglomerate, consisting of and depending upon a lot of things, including the pure enjoyment of doing research mathematics. To put it simply, if I were to swear-off mathematics, I would be unhappy, and this unhappiness would taint the rest of the conglomerate, regardless of how much time I saved in the process.

[P] H. Poincaré, Mathematical creation, The Creative Process (Brewster Ghiselin, ed.), Mentor Books, New York, 1952.

[B] D. Barry, Great moments in science, Chicago Tribune Sunday Magazine, March 30, 1997.

Getting Started in Research:
A personal perspective
by Frank Sottile

In the 19 February 1997 issue of Concerns of Young Mathematicians, Kevin Knudson asked how do you make the transition from graduate student to budding research mathematician? That is, having worked on a problem for several years with your advisor's attention and counsel, how do you start independently seeking out and solving your own problems? Anecdotal evidence and advice I have been given point to this as being a critical transition, one which many (most?) young mathematicians do not make.

I don't have the answer. Mathematical research is a personal pursuit, and what works for one may not work for another. However, I think I have made that transition and have some observations. I will mix these observations with the story of the first year of my career, so this won't apply to everyone, but hopefully it will be helpful to some.

I managed to prove a new result within two months of starting my first job. While I was fortunate to get so quickly started in new directions, I feel that it was absolutely necessary; you can't go very far extending your thesis or living off ideas that your advisor sends your way. The key is to generate your own ideas. One way to do this is to attend lots of talks and conferences and talk mathematics with as many people as possible. Continually ask yourself questions. I recommend casting a wide net for potential problems; not only are you more likely to find something to your liking, but you will learn more along the way. You don't have to work on problems similar to your thesis for the rest of your career. That said, you should publish the main results from your thesis. I didn't do this for about a year, and the delay later cost me a lot of time and grief.

I had heard of people who got their Ph.D., moved to a new city for their first job, then their first term started and they got caught up in their new responsibilities. Almost before they knew it, several months had passed during which they hadn't done any research and then they found it difficult to get started.

I was determined to avoid that fate. As soon as I arrived and set up my household, I started to think about two ideas I had earlier in my graduate career, but were unrelated to my thesis. These ideas were vague, and I didn't get far with them. One was cut short by the appearance of a manuscript—by someone with whom I had shared my partial understanding the previous spring. I still think the other is interesting, but school quickly started, and when I had the time again, I had other ideas.

The first week of any academic term is a time to organize your classes and your schedule and to see your colleagues after a break. It isn't a good time for research. In a new job, there is more dislocation: adjusting to new responsibilities, a new town, new department, and new people. In particular, you need to establish a rapport with your students and find a good work schedule to follow during the term.

It took me several weeks before I was able to do much else, and by then I had other responsibilities: I was scheduled to give a seminar at my home institution, and then speak at a regional conference the next weekend.

Giving and attending talks is an important component of my research activity. I use talks, particularly at my home institution, to help me organize my ideas or cast them in a new light. It is a challenge to present a coherent explanation of my work in any format. The choices this forces me to make always lead to improvements in how I think about my work. Giving a talk introduces me to my audience, and such introductions lead to worthwhile interactions. This is also a reason to attend seminars and colloquia, and to travel to conferences. I foot the bill for some of my travel if no outside support is available. After all, these interaction are a vital part of the mathematical enterprise. Also, a weekend conference is a very welcome and refreshing break from the day-to-day business of a typical term.

Another responsibility I had was a grant application, which turned out to be a good use of my time. In that application, I tried to convince other mathematicians that my ideas were worth supporting. That exercise forced me to generate new ideas of what to work on. To assist my thinking, I contacted several people at other institutions and started a dialogue. I have recently written two joint papers with people I first contacted at that time, almost 30 months ago. The value of those contacts isn't just these recent papers, but how they helped my thinking about mathematics to evolve.

How does one make professional contacts as a fresh Ph.D. recipient? I already mentioned giving talks and attending conferences. When I travel or when the department has visitors, I socialize and ask questions that come to mind. It is easiest to meet people who live nearby. I met one of my recent coauthors when I learned that he had also just moved to the same town.

One way I have met other mathematicians is by distributing copies of my papers. When I complete a manuscript and get ready to send it to a journal, I make a list of other mathematicians who may be interested in these results and send them copies with a cover letter describing the main results and why they might be interested. This list consists of people whose papers I have read or cited, people who I have heard speak or heard about, people who have asked me for copies of my work, and my local colleagues. I am not shy about this; at worst, they won't read my paper. I also put my papers in electronic preprint archives. I have met several people this way, some of whom have had me visit or give a talk and one with whom I just wrote a paper.

When writing that grant application, one person I contacted pointed me to a conjecture (in algebra) in one of her papers. At the time I finished the grant application, I hadn't done any new work in the several months since I handed in my dissertation. Rather than get started, I traveled to the Midwest to give a seminar, visit a friend, and attend a weekend conference. While there, I had a flash of insight one morning in the shower (no kidding). I realized how it should be possible to prove the afore-mentioned conjecture using geometry.

Generating ideas is necessary, but then you need to bring them to fruition. Research, like other quality pursuits, requires quality time. Discipline may be the most important part of my doing mathematics. I impose deadlines on myself, reserve time solely for research, and try to use my time efficiently. I jealously guard my research time; otherwise it evaporates as there are many other demands on my time. In short, hard, dedicated work is essential.

When I returned home after that trip, I arranged my schedule to give myself some uninterrupted blocks of time. This was a matter of changing when I did my class preparation and other necessary tasks (email, meeting with my TA's, grading homework, lunch, *etc.*) so that I would have these periods. Then I used this time to try and prove this conjecture. After a very dedicated few weeks, I did it, and I felt great. After a couple of hours of euphoria, I began the hard part—writing up this result and its obvious extensions. I spend more time writing than research, and writing is not nearly as fun. It is absolutely necessary to spend this time—if my papers are not carefully written and if I do not make every effort to improve the exposition and clean up the arguments, then few will read my papers. Remember, the readership of any paper is a monotone decreasing function of its line number.

While writing up this result, I started to work with a coauthor on a further extension, which would have been a fantastic result. After six months of my doing nothing else, we gave up. This failure to get results is common, many promising lines of my research fizzle. When this happens, don't let it keep you from trying. When working on a problem, you inevitably learn some new mathematics or get some ideas, or obtain partial results.

Partial results are not necessarily second-class; the problem might be harder than you thought. If the partial results are interesting and lead somewhere, then they are publishable. Only last summer, my coauthor and I came back to this work and wrote a paper whose genesis was the successes we had among our many failures. We submitted this to a good journal which accepted it quite quickly. I think this is the best paper I have written, and we have several related papers planned. While we have yet to solve the original problem, this has become my most interesting line of research.

During the last half of that first year, I did other things as well. I attended an instructional conference in an area that I am only now just beginning to work in. There I met someone who was very enthusiastic about my thesis and gave me the encouragement I needed to write it up, and who has now become an important mentor. I also met a student there whose research gave me an idea of an application of some of my work. Now we are writing a joint paper on those applications.

At the end of my first year out, while I had failed even to massage my thesis into a preprint and had only solved one new problem, I think that I was on the right track: I had made serious work a habit, I was in regular contact with a number of other mathematicians, and I was thinking about a number of different problems.

And Then the Students
Knock on Your Door ...
by Evelyn Hart

I managed to get some research done while I was at a school that emphasized teaching yet still expected some research. I forced myself to hide from my students at times. Some people say to save an hour per day for your research and not to cancel it unless you would cancel a class. That's the right attitude, but I had trouble saying no to students. I was able to get a faculty study in the library (close to my office, but no phone, no email, no one knew I was there), and I ran there whenever I could. I even once made an appointment with myself and wrote it in my calendar. When a student asked to meet with me at that time, I said that no, I had a meeting then and could we find another time to meet? Of course we were able to meet the next day. I never would have had that hour to myself if I had not written it in.

I've found that students say, "Meeting with you this afternoon is impossible. How about tomorrow?" It turns out that "impossible" means that a meeting then would be inconvenient. I push for times that work well in my own schedule.

I accomplish a lot in the morning if I can get out of bed. No one calls or interrupts between 6 and 8 am. I can spend the rest of the day feeling less frantic, because I've already put in time on my research.

Students usually have no idea that we have to do research. I tell them that unless I do research I will lose my job, so could we please not meet on Tuesdays. I don't get to spend all of Tuesday on research, but at least I have one day with no meetings with students. I find that they are surprised and then supportive of my needing time to myself.

Also, any time a student asks for an appointment for help with homework, I reserve the right to have another student from the same class come at that time. If a student wants to talk about a grade or something private, the student needs to tell me that. They do deserve to have access to me, but I am not a private tutor.

Be sure to start research early in your time at your first job. Some referees take a year to evaluate a paper, so when I moved to a school where research is more important and was told it would be good to have 5 papers in 5 years, I thought of that as 5 papers in 4 years. Thank goodness I have had very good luck with response times from referees. But I know of many people who have not. Papers always take longer to write than I think they will, and then there are usually revisions needed after the referee sees the paper.

During summers and winter breaks I have always forced myself to put off preparing for the next semester until the last possible moment. I'm sure my teaching has suffered, but my teaching seemed to be acceptable and the research was what I needed to improve. After tenure is the time to set one's own priorities for the balance between teaching and research.

Time to stop giving advice and start following it . . .

A Research Mentor is a Good Thing to Have
(and Other Advice)
by Curtis D. Bennett

If you are fortunate enough to find yourself at an institution with senior researchers in your field, it is very helpful to find a mentor or mentors to communicate with. Ideally mentors should not take the place of your Ph.D. advisor. They should be individuals with whom you can comfortably discuss your mathematical ideas, even if they are not fully developed. Unlike your Ph.D. advisor however, you should be forming a more collegial relationship with them. Mentors don't necessarily have to fully understand all the background of what you are talking about; in some ways, their most important role is to serve as a sounding board for ideas. After all, the process of explaining to someone else what you are trying to do will help you understand it better. The other role of a mentor is to provide you with someone to better show you the ropes. You should be able to get suggestions from a mentor on publishing your papers, giving talks, or even applying for jobs. Remember, the best resource for a mathematician is other mathematicians.

Even if you are fortunate enough to find a mentor, it is crucial that you keep in contact with your advisor. Your advisor is the one person besides yourself who best understands your research. Advisors can provide you with insight about what topics are currently of interest, and they can also tell you what articles to read and where to look for your next paper. With e-mail, keeping up this contact is relatively easy, and you certainly should do so.

If the school you are at has seminar talks in your area, you should be sure to attend and give talks in the seminar. In particular, it is worthwhile, and frequently expected, for you to give a series of talks on your thesis and subsequent research. The advantages of this are many. First it gives you more experience at giving talks. Second, you probably haven't had much of a chance to explain your research to people who are less versed in it than your advisor. Even if you have run similar seminars as a graduate student, odds are that your advisor was in the audience then. This may be your first chance of working without a net. Another advantage is that talking will force you to become more of a part of the department. It is too easy when working at a part time position to fall into the cracks of the department. Since it is important for you to develop a relationship with mathematicians in your field, it is crucial for you to be active in your department if you hope to continue your research.

When attending seminar talks given by colleagues, try to take 15 minutes to an hour afterwards to think about what they have said. Perhaps you can come up with a small improvement of their result, or perhaps you will have some deeper questions about the talk. I was given this advice four years ago, and I consider it one of the best suggestions that I've ever been given. On one occasion it has led to a joint paper for me. One important comment here; the purpose of this is to help broaden your knowledge and to help you make new connections. Thus, the real reason to think about what was said is to firmly cement the ideas into your head.

Conferences are another good place to find people to talk to about your research. I think that it is best to give at least one talk a year at some conference. Talking will introduce you to other mathematicians doing research in your field, and it will let you know what others are doing. Depending on your field, it may happen that someone else is doing almost exactly the same thing you are. This is only a problem if you don't know about it. Otherwise, you can compare notes and collaborate. This will be very helpful in the long run.

Perhaps the hardest thing about the above suggestion is finding the right conference to talk at. This is where your advisor or mentor should be able to help out. If for some reason this doesn't work, find mathematical siblings and ask them for advice. You should also look at the organizers and titles for the AMS special sessions. If you recognize any names, contact them and find out if it would be appropriate for you to talk in their session. They may say no, but if they have room and your talk is related to their session, they will probably be happy to include you.

Keeping Your Research Alive
by Julian Fleron, Paul D. Humke, Lew Lefton,
Terri Lindquester, and Margaret Murray

The discussion below is a transcription (by Julian Fleron) of a Project NExT panel discussion which took place at the 1995 Joint Mathematical Meetings.

Introduction: by Humke. It feels good to discuss a topic which I have thought a great deal about and for which I have a great affection for. As if it were yesterday (now almost twenty-five years ago!) I remember the transition from graduate student to faculty member, the stunning new demands on my time, the responsibility for curricular matters I'd never thought about, the energy level necessary to do my job. The foremost question I had about my *professional career* was whether I would even have one! And I wanted one!!

Before introducing the panelists I'll give a definition (for local use only) and make three observations.

By *professional career* I mean my niche within the community of professional mathematicians. Specifically I'd like to put a bit of distance between what we're discussing today and my role as classroom teacher. For the record, I believe that classroom teaching is very professional and very important, I love it. But today's discussion is focused on the role of a mathematician outside the classroom.

During my first few years this meant my niche within the community of research real analysts and it still has that meaning for me. But my niche has grown to mean much more, and I'll discuss that growth in my remarks below.

For me, teaching and research are very different worlds. Different types of demands, different needs, different environments, and indeed different colleagues. And I live in both worlds, I have great enthusiasm for both worlds and I think that this situation is "OK". So in my St. Olaf way, that is my first point:

> Live in two worlds, but be of one spirit.

At one of the recent NSF presentations, the moderator issued the "fact" that to be effective, a department of mathematics needn't have *teacher-scholars*, but merely have some teachers and some scholars. Don't believe it! My own opinion is that:

> The *teacher-scholar* model is far more dynamic and effective.

This is because by living in two worlds you become the link between the excitement of cutting edge mathematics and your students. It is through you that mathematics comes alive, not just the mathematics of one hundred years ago but the mathematics of today and the mathematics of the future. It is precisely because you embody that link between past and future mathematics that your students can see themselves becoming part of the mathematical enterprise. This is one reason why a lively professional career is important. (It can also help with getting tenure!)

My last point is that:

> The program I've outlined above is manifestly impossible to attain!

On the other hand, you can make a good approximation and have a great deal of fun doing it. Set yourself some high goals, but be tolerant of yourself and most important, have a good time with what you do.

Lew's Liftoff. There are several good reasons for keeping your research program alive. An obvious one, for those who are fortunate enough to be on the tenure track, is to assure a sufficient publication record to be granted tenure at your institution. Of course, if you are not in a tenure track position, you should have just as much if not more incentive for working on your research. After all, you want to remain a strong candidate for when you are back in the job market.

Even if your duties and interests are primarily in the teaching of mathematics, there are still compelling reasons for continuing to remain active in research. It's important for our students to realize that mathematics research is a current and ongoing activity! As working mathematicians, we have a responsibility to communicate the vibrancy of our subject. You may know that new results and ideas are being discovered every day, but do your students? Finally, there is also the general benefit that research, like any other mental exercise, helps keep your mind sharp. Of course, since it's safe to assume that you are reading this article, you probably don't need convincing that research is important. The question is not "why" but "how". I can't offer a definitive answer here, only some suggestions.

1. Keep in touch with the experts in your area of interest.

This may mean simply writing a letter or email to ask for relevant preprints. You have to get your name out there. Go to conferences and present your work. I recommend the smaller regional research conferences as opposed to the large annual meetings. Not that the annual meetings are bad, in fact, if there's a special session in your area it can be quite productive. But smaller conferences often have fewer distractions and more opportunities to interact with people who have similar interests. Even if you don't know anyone, go to listen and ask questions and learn. Once you attend a few such conferences, you will start to see several familiar faces and you will have begun the important task of establishing professional contacts in your area. (Remember, you will need references for upcoming tenure and promotion reviews and any future job hunts, and you can only rely on your graduate school faculty for so long!) Don't be intimidated, most active research groups are happy to welcome new people and they may well suggest some interesting open questions for you to work on.

2. Work within your own institution.

Ask your colloquium committee chair to invite someone in your area. Start a seminar, you only need two or three people with common interests. Don't let that feeling of isolation defeat you. Even if you're at the University of the Moon and you need to get together with colleagues at the Asteroid Belt Community College, make time to do it. It's worth it in the long run.

3. Make your research a priority.

Set aside particular times of the week and find a quiet place to go where you'll be free from interruptions. This is really important, but it can be very hard to do. I don't want to turn this into a piece on routine time management techniques so I'll cop out by saying consult your local library for general help with the basics.

One thing to do if you're having trouble managing your time is to say "no" when you are asked to do additional time consuming projects. There are many worthwhile projects but you can only do one thing at a time. In the early years of your career, you need to establish a research record. If saying "no" isn't an option, try to negotiate some release time from your teaching so that your research doesn't suffer by default.

The main thing to remember is to maintain a high level of professionalism in all of your projects, whether research, teaching or other professional activities. This translates roughly to "Do a good job, and make your mama and papa proud." If you consistently do good work, then you will gain respect among your peers no matter what direction you pursue in this rapidly changing profession of ours.

Marge's Musings. I received my Ph.D. from Yale in 1983, and came out onto one of the best job markets in mathematics in the past twenty-five years, so it is with some trepidation that I offer advice on how get your own careers going and keep them afloat. Let me offer a few caveats: first, the job market was good when I finished my degree, but it hadn't been so good just a few years before, so mentally I had prepared myself for the worst. Second, although I had a number of job offers to choose from, I somewhat deliberately chose a job situation for myself that offered some of the same challenges that you are facing in your careers right now.

I chose to go to Virginia Tech, a university somewhat off the beaten track, where I would be the only person working in my area of research. As I have always taken a somewhat iconoclastic approach to my life and work, the projects I decided to pursue were somewhat at odds with what was expected of me. For example, as an untenured faculty member I decided to begin work on a book project. In hindsight I am very glad that I did this, but the rewards of this sort of project were not immediately evident. In a profession that puts a premium on timeliness, choosing to do this was definitely a risk that put my career on the line.

I recognized that my mathematical survival depended on making others aware of my work. I attended as many conferences, and volunteered to give as many talks, as I could possibly manage. I applied early and often for NSF grant support, and it was my great good fortune to be successful in landing grants. The result of these efforts is that I was able to communicate on a regular basis with other mathematicians about the long-term project I had undertaken, and I was able to convince a good many people of its significance. The moral of the story: choose your projects with courage, and make every effort to advertise and promote your work.

All that is in my past. As for the present—yours and mine—let me say that these are very confusing and tumultuous times for the entire mathematical community. The working environment for *all* mathematicians, young and old, is up for grabs, as the community is in the process of redefining itself. This is a time of trouble, but also of opportunity.

Though the mathematical community has historically been rather rigid, this is a time of unprecedented freedom to undertake whatever sort of meaningful scholarly work in mathematics that captures your interest.

Your greatest assets are creativity, flexibility, adaptability, and the willingness to take an off-center approach. Whatever you decide to undertake, the most important thing is to avoid professional isolation.

Promote your work, seek allies, enlist colleagues, speak in seminars and at conferences, seek out new connections to mathematicians and to professionals in other disciplines.

It is quite unlikely that you will lead the same kind of professional lives that your professors did, but this should not be reason for despair.

You need always to remember that you have unusual training and skills. The world—both inside and outside of mathematics—is waiting for you, full of problems to be solved.

Terri's Tenacious Techniques. Rhodes College is a highly selective, private liberal arts college with a student population of about 1400. The teaching load is three courses per semester which is equivalent to nine hours per week in the classroom. Normally, because the students are good, I can expect a lot from them. But in turn, they expect a lot from me. I am expected to be available to students outside of the classroom to answer their questions, help solidify concepts, or just talk about mathematics. I spend about eight hours per week meeting with students outside of class.

The service component of my job consists of plenty of college-wide committee work, special college projects (such as devising a new registration system), departmental curricular projects, and advising. (And actually advising is one of my most time-consuming endeavors.) And of course there is a research expectation.

As a graduate student my prevailing thought was "if I can just find a good problem," or "if I can just find a way to prove this conjecture." As an assistant professor, my prevailing thought was "if I can just find the time!" Since professional activity is a continuous process of growth and change, we have to find a way to make time for it all along the way so that this professional development is rich and meaningful, rather than disjointed and incomplete.

The first suggestion may sound simple or even trivial, but being successful at it has probably helped me more than anything to create time for professional activities: Reevaluate how you organize and structure your classes to see if there are any ways to achieve what you want to achieve more efficiently. For example, clarify your expectations to your students in your classes verbally and in your syllabus so that students know exactly what their responsibilities are, and so that you know exactly what your schedule will be for the semester. For instance, set your office hours in your syllabus on the first day of class, and try to encourage students to stick to these times rather than allowing them to wander into your office at any time during the week. This will help you organize your grading and testing and enable you to set aside big blocks of time for your research. Think carefully about your testing schedule for the semester early on, so you don't spend time creating and administering make-up exams and quizzes during the term. Assign projects and writing assignments to groups of students—not only is this a wonderful way for students to engage in problem solving with their peers, but it cuts down on grading.

Encourage your department to organize and support student-run help sessions to be held at night to supplement the help you provide during office hours. Moreover, when it is appropriate, request multiple, back-to-back sections of courses so that you have fewer preparations and you can give the same exams for both classes.

In your upper level courses, try using more original sources instead of textbooks. Get students involved in critically reading papers and giving presentations. These things, though seemingly small, will keep you in a more active mode of inquiry and possibly lead students in that same direction. This way, even if the papers are not specifically in your area of research, you are learning and investigating in the manner in which you investigated as a graduate student.

In service, since I am presently the only tenured woman in the Natural Sciences Division in my institution, I have been asked to serve on college committees where gender and divisional diversity are desired. Sometimes I just have to say no. Certainly, as an untenured faculty member this is harder to do—but it's a necessary response sometimes in order to maintain a legitimate professional activities calendar. If you can't say no, or have a project offered to you that you would like to tackle, turn it in to a professional activity! (For example, maybe the curricular reform project you've been assigned lends itself to writing a grant proposal to carry out such a reform.) There may be ways to alter your service duties a bit so that you can gain in other ways by disseminating the ideas you have used.

In research, at least in the beginning, it is important to maintain your collaborations. Keep in touch with individuals with whom you have worked so that talking and thinking about problems doesn't become an unnatural thing to do. They can also help you to stay current with the literature in your field. Attend seminars and meetings as often as possible—that interaction is as valuable as any.

And finally, as a new Ph.D., try to focus in one direction professionally. Once you have established yourself solidly in one area, then you will be in a better position to dabble in other areas of interest later on. Think about what you want to be doing in the next five years and set your sights there!

Paul's Prattle. In 1978, Fred Gehring of the University of Michigan surveyed *all* of the Michigan mathematics Ph.D.s. (I'm quoting this survey from memory; warning enough for those who know me.) Among the things that this survey showed was that less than half of the survey group ever published anything (even their own dissertation work); of those that published something, less than half published more than one paper beyond their dissertation work; but all of those who published at least five papers continued to have a lengthy and rich publishing record (twenty or more publications). This says to me that the first few years after graduate school are a critical period in the professional life of a mathematician and that creating a *beachhead* in the publishing world is both necessary and sufficient for a generous professional career.

Below I'll make a few brief comments about my own experience and then end with three specific suggestions.

My **Early Career** was characterized by traditional publication work: learning, reading papers, thinking, writing up results, submitting papers, revising papers, learning to use *Typits*.

My own career began at Western Illinois University with the usual stuff. I published *most* of my dissertation results and began attending research conferences in real analysis. At these conferences I met many of the researchers who would become my fast friends and frequent collaborators.

The period when I was **Five Years Out** was more of the above, but with some important additions: refereeing, editing, symposia organizing, grant writing. I also learned to use the IBM *Symbol Ball*. In addition I worked on my first two "applied" problems; one a design of plowshares for John Deere and the second some Operations Research.

My research in real analysis continued and "applied" problems became a periodic part of my life. I began attending the weekly *Real Analysis Seminar* at the University of Minnesota, something I've continued ever since. But by the time I was **Ten Years Out**, there was something quite new, curriculum reform. After my move to St. Olaf I became involved with the calculus reform ideas, ran conferences, discussed the issues with Mike Evans and Jerry Uhl (two other ancient real analysis colleagues), learned SMP (a precursor to Mathematica) and Maple, worked on several calculus reform committees, wrote grants by the score and became acquainted with a new and enthusiastic crowd. I also "learned" *TROFF-NROFF-EQN* and TEX.

And so it goes; I continue to be active in real analysis research. Editing takes much more time now that I serve as a Managing Editor of the *Real Analysis Exchange* and as a consequence, I have about shut out refereeing and reviewing. I'm not as involved in calculus reform as I once was, but I still maintain an active interest. I have now "learned" LaTEX and use it almost exclusively for technical work.

If I had some hints for others, what would they be? I think these.

Humke's Hints. Set aside time weekly for professional work. There are those who have claimed this won't work, but it works for me. I treat my research time the same way I treat my class time. It's high priority and I don't cancel my research time unless I would cancel a class for the same reason.

Keep something on the burner. If you have a problem, you can work on it whenever you have a spare minute. In my experience, research takes some long periods of concentrated work, but it helps a great deal to have some aspect of your problem to think about when you have a free minute or two, when the party becomes dull, or your lunch date fails to show up or .... My friend Gyuri Petruska once said that

> when I was young I'd spend five minutes "warming up" and then I was ready to work. When I was a bit older I had to spend twenty minutes warming up before I was back into my work. Now it seems I spend all my time with warm ups!

For Gyuri this is **not** true at all, but there is truth in his very Hungarian view of himself. If you *keep something on the burner* you'll minimize your warm up time.

Attend the small research oriented conferences in your area. Small conferences are attended by those who are active in your area. You'll get to know them and become acquainted with the lore of your discipline. You'll also discover the topics of current interest and some open problems; that's always helpful. Trivial problems and impossible problems are both a dime a dozen; good problems are more difficult to find. I find I'm always "fired up" by a real analysis conference and that motivation often translates into a result or two.

Questions, Comments, Remarks. The following is a loose transcript of the question and answer period. What appear below are not direct quotes, but they are closer to quotes than they are summaries. The letters *P*, *T*, *M*, and *L* stand for Paul (Humke), Terri (Lindquester), Marge (Murray), and Lew (Lefton), the four panelists.

[Q How does one deal with the spouse/kid issue?

[L] Anybody with a small child doesn't sleep. My daughter has taught me how to say "no." I have shut a lot of things off.

[M] I have found there is a perception that men shouldn't claim time for their kids. It's kind of reverse sexism.

[P] For me, family is the top priority.

[Q] I have a question about grant writing. Some of us would like time to work on common research problems together. Is it reasonable to write a grant to bring people together to do this? Is money available for this sort of thing?

[M] You should start at your home institutions. They often have travel grants for young faculty.

[P] A group of real analysts had the same question 20 years ago. Their efforts have blossomed into an annual summer symposium held at various universities. They arrange to stay at dorms which only costs $10 a night or so. Also, the NSF funds young people; it has "affirmative action by location."

[T] The AWM often has travel money available.

You could get together at either end of a conference related to your research interests. You should also consider more obscure places. School faculty development grants might support such requests. Rotational grants for smaller amounts of money get rotated around. You should check with the state for educational sources. I organized a regional graph theory conference that was funded by the Office of Naval Research. There are also PEW grants; grants for scholars to come and visit your institutions. Additionally, conferences for undergraduates usually have lots of information about grant writing.

(As a specific example of "keeping your research alive," and the priority this must take, Lew Lefton leaves at this point in the discussion to give a research talk at an AMS session.)

[Q] What about grant possibilities for a reduction in load so you can have time to do research?

[P] There are some outside funds for this, but you can't expect your school to pay for all professional needs. Perhaps there will be a local program to fund reduction in load, but, for the most part, the responsibility for finding research time will fall to you.

[M] You should check your school's development offices. And their sponsored funds offices.

[Q] I just feel overwhelmed. Talking about pedagogical issues, how do you do this together with research? It's not a matter of just being able to say "no." I feel uncomfortable trying to redefine the job in a visiting or non-tenure position.

[Q] (Continued by a different questioner.) None of us have come up for tenure. We are in weird positions. Even if the department supports pedagogy, final tenure decisions are not made just in the department. They might be able to tell us what they want in terms of what they've had in the past, but that is a very different situation than now. It's almost as if we are the guinea pigs, coming of age during this time of great change in the mathematics community. We are scared how our tenure decisions and careers might be affected by this.

[P] Some things never change. The kind of professional development we are talking about is important to you for your careers independent of the current state of the mathematics community. It would be bad for you and for the profession to ignore this type of professional development. It should be a priority. I don't mean traditional research necessarily, but I also don't mean just education reform. You can spend all your time doing curriculum development. I will say that you do need to be well aware of all the policy documents that apply to you in your department.

[T] It comes down to analyzing what you are doing. Are you doing it efficiently? Think things through on the front end, even regarding pedagogical issues. For example, how you set up exam procedures, office hours, and so forth. Organizing them effectively can save you a great deal of time.

[P] You should also be sure to communicate your thoughts on these things to the people around you. In particular, by communicating alternative, more efficient means of getting some of the shared tasks accomplished.

[T] Professional development is institution specific, but much of what we are talking about is universal.

[M] I tended to work on big long term projects. I was taking a very big risk in doing things this way. Because of this, it was important for me to talk to people about what I was doing. You need to make sure to portray your activities in a positive way. I could have written a book with a closed office door, but I wouldn't have gotten tenure that way.

[P] There are people with national stature who know how much time all these forms of professional development take. It would be appropriate for them to address these issues. So you must be careful to keep track of what you do and who might be able to comment on your efforts.

CHAPTER 5

What to Do with Your Research
Once You've Done It

The previous chapter discussed how to find the time and the resources to do research; this chapter discusses the next stage. Since
our research is done within a community of mathematicians who
are often investigating similar—if not exactly the same—problems,
the first article addresses some of the ethical considerations of
working and communicating with other mathematicians. The last
three articles in the chapter offer advice about two common concerns: writing mathematics so that it will be suitable for publication, and choosing an appropriate journal.

Ethical Questions on Discussing Research
by Todd Wilson, Charles Holland, Steven
Krantz, Richard Phillips, and Ronald Solomon

The following article is compiled from a virtual panel discussion held on *Co YM*. The
whole idea for the discussion was prompted by a request from Todd Wilson to clarify
the mathematical community's ideas on "the ethics of mathematical practice". He
asked three questions in particular:

1. How should one discuss preliminary results or research ideas with other
   researchers? What about the danger of these researchers following up on
   your idea?
2. Joint papers never have exactly equal contributions by all authors. It seems,
   therefore, that some authors will have something to gain, while others will
   have something to lose. To what extent is this true?
3. What should you do if you are refereeing a paper that contains results
   that you also discovered and were planning to publish, especially when the
   paper's results are slightly weaker than your own? How generous should an
   anonymous referee be?

Charles Holland is a former department chair of Bowling Green State Uni-
versity, editor of three books, and has served as a referee for 35 years. Steven
Krantz has served as editor of five journals including the *Notices of the AMS*, the
Mathematical Monthly, and the *Journal of Mathematical Analysis and Applications*.
Richard Phillips is a former department chair of Michigan State University, former
chair (and associate chair) of the Michigan Sections of the MAA, and a member of
the editorial board of the *Journal of Group Theory*. Ronald Solomon of Ohio State
University is a former editor of the *Proceedings* of the AMS.

Their responses are below.

1: Discussing Preliminary Research. *How should one discuss preliminary results or research ideas with other researchers? What about the danger of these researchers following up on your idea?*

Charles Holland. My view is that in the long run it is much more interesting, and even more profitable, to share, rather than to protect some small piece of turf.

The vast majority of mathematicians are scrupulously ethical in such matters. They give credit where it is due, identify sources of problems, any known work previously done, any help given during discussions, etc. Most even avoid, or at least hold back on problems a student is known to be working on, and on problems that are discovered in the course of refereeing a research proposal. So the probability of getting mugged is very small. And if it does happen, then you know to be careful in dealing with *that* person in the future. A slightly different scenario arises, however, because it really is difficult to avoid working on interesting problems that you become aware of. Thus, any thing you make public should be considered fair game. So, if you are very close to solving some problem of general interest, and it is important to you to get individual credit for it, it is OK to be close mouthed for a while. However, it wouldn't be right to let others slave away for a very long time without letting them know what progress you had made. Here are some specific scenarios, and their typical resolutions: Mathematician A has made some progress an a problem, discusses it with others, then mathematician B solves it. (1) If A had most of the solution and B only supplied the final step, then A publishes, crediting the final step to B. (2) If both A and B supplied major portions of the solution, a joint paper is in order. (3) If most of the major work was B's, B publishes and credits A with the origin of the problem and perhaps some portion of the solution. The distinction between (1),(2), and (3) is determined by the parties involved.

Steven Krantz. Ideally, one should be able to do as Paul Halmos advises: "Cast your bread upon the waters and see what returns." In plain English: talk to everyone about everything. The benefits accrued are worth the risk. For the most part, this is the course that I steer. However there are certain individuals whom I categorically will not trust.

The unwritten rule is that if someone, especially a younger mathematician, comes to you and says "I'm working on so and so" and then asks some questions, then you are supposed to leave your hands off his question. If you are a senior guy and a young person approaches you in this way and you know how to do the problem you are supposed to be very gentle. It would be extremely rude and discouraging for the senior guy to say "That's trivial and here is how you do it." Better to say "I think that there are some related ideas in thus and such a place. Have a look."

The rules, if there are such, are different when a senior guy says in public "I'm working on so and so and here is where I'm stuck". That, in effect, is to be interpreted as a challenge. If some young turk comes back and says "I have an idea" then the senior guy can say "Let's write a joint paper" or "that's very nice, why don't you submit it to my journal."

I'm beating all around the bush here, but the main point is that you can trust most people most of the time, and you'll benefit from being open. But you will learn quickly that there are certain individuals you should stay away from. And "Once burned, twice shy", as they say.

2: Contributing to Joint Papers. *Joint papers never have exactly equal contributions by all authors. It seems, therefore, that some authors will have something to gain, while others will have something to lose. To what extent is this true?*

Richard Phillips. Professional decorum and evaluation of joint papers is a real blur. My own view is that if there are multiple names on a manuscript, then all those people made a contribution. I realize this is not always true, but it's a fairly safe working hypothesis. How joint work is viewed when it comes to evaluation (= $$ and promotion and tenure) runs the full range. There are very good mathematicians who write only joint papers and there are others that perpetually go it alone. In general, I think that if there is a junior/senior tandem and the senior person has name recognition, it probably benefits the junior author (of course, some are going to say that Joe Senior had all of the ideas, *etc.*, but this is generally a minority view). Another aspect of joint work that should not be overlooked is that such work indicates the parties involved can actually work with other people—this is a plus. Overall, I don't really see how anyone loses in a joint piece of work. A much more complicated issue is when co-workers decide whether or not a piece of work should actually be authored by more than one person.

Charles Holland. It is hard to see how anyone loses in a joint paper. No one is forced to be a joint author. It is probably true that the contributions are never equal, in some sense, but it is also true that the paper is uniquely the product of those particular authors—it wouldn't have been the same without any one of them. Most joint-authored mathematical papers list the authors alphabetically, precisely to diffuse the question of who contributed most; and most joint authors won't discuss that question.

Steven Krantz. Now for joint papers: read what Paul Halmos says in his memoir. Ideally, once you've started on a joint venture everyone is a co-author until the end. It is a given that no two co-authors will make just the same contribution. What one hopes is that the whole will be greater than the sum of its parts. That is, in the best of situations each author contributes something that the others could not have.

Of course situations can and will arise where it is clear that one author is contributing nothing. This can happen even with the best intentions of that errant author. The other author(s) could be leaving him in the dust. The gentlemanly (pardon the sexism) thing to do in that circumstance is for the odd man out to offer to withdraw. At the very least the odd author could work like hell to digest the material and help to write the stuff up.

What typically happens is that, without anything being explicitly spoken, this one author hasn't lived up to his part of the contract and no future collaborations will occur.

Of course I'm painting an ideal picture here. A lot of nasty fights have been caused by attempted collaborations. I have even witnessed one which involved guns, lawyers, and death threats. But my experience has been almost uniformly favorable. The five percent risk is greatly counterbalanced by the 95% certainty of profitable experience.

3: Refereeing a Paper with Familiar Results. *What should you do if you are refereeing a paper that contains results that you also discovered and were* *planning to publish, especially when the paper's results are slightly weaker than your own? How generous should an anonymous referee be?*

Charles Holland. Have you not yet submitted your results? Then the paper you are looking at has priority and should be accepted. However, if your more general result would substantially improve the paper, you might suggest (through the editor) making it a joint paper, in which case another anonymous referee might be brought in. Acceptance of the proposal is up to the author of the paper you are looking at. Even if your paper were already submitted, a proposal for a joint paper might be in order. Of course, sometimes the methods are so different that publication of both solutions is appropriate.

Ronald Solomon. My philosophy would be this:
Case 1: The results are essentially identical, as are the proofs.
Subcase a: Your paper has already been accepted by a journal. It's up to your sense of fairness. Either reject the paper on the grounds that the result is known and due to be published in ... , or contact the author and offer to add his name as a co-author of your paper.
Subcase b: Your paper is at a similar stage of the refereeing process. Contact the author and propose that you resubmit the work as a joint paper to be refereed by a third party.
Subcase c: You have not yet submitted your work. Throw yourself on the mercy of the author. Let him know and propose a joint paper.
Case 2: Your results are slightly, but only slightly stronger ... or they are the same, but you have a better proof (at least in your judgement).
Subcase a as above: Reject the paper: A better result by (your name) will appear shortly in (the journal).
Subcase b: Follow either subcase a or subcase c strategy
Subcase c: Contact the editor. Ask him or her to contact the author and say: The referee informs me that he recently obtained similar but stronger results. He or she (*i.e.*, you) would be happy to publish this as a joint paper with you. If the author declines the invitation, tell the editor that you recommend against publication on the grounds that you intend to submit a better paper shortly. If the editor wishes to contact another referee, the editor is of course free to do so.
Case 3: Your results are significantly better.
Reject the paper: Much better results in this vein have been obtained recently by (you) and will appear shortly.
[Again, the editor always has the right to overrule you, seek other advice, *etc.*]

Steven Krantz. I am concerned about some of the advice given above. I know of a circumstance where a guy rejected a paper and then contacted the author and suggested that they write a joint paper. When he was found out, he was made to wear the scarlet letter for life. No kidding. Now nobody suggested this in your newsletter, but one must be careful.

Of course the example I just cited is extreme, and none of the writers above suggested this sort of behavior. I think that a referee should under no circumstances contact the author of a paper he/she is refereeing. One always has the option of contacting the editor, but my guess is that one won't get much satisfaction since most editors won't know what to do and/or won't want to be bothered. Certainly most editors won't want to get involved in a situation of suggesting to the guy who submitted the paper that he make it a joint paper.

It sounds like a Mandarin system, but the way this has often been handled in the past is that the person doing the refereeing approaches some senior guy—a friend—and asks him to act as go-between. One of the shortcomings of the system we have is that we don't have a system.

I'm sorry that I cannot offer a more cut and dried solution for this circumstance. The main point I want to make is that this is a tricky situation. Be careful.

(Re)writing a Thesis, and other Mathematics, for Publication
by Curtis D. Bennett

After I finished my thesis, I had gotten burnt out on the topic. This is not uncommon. It is fine to feel this way, but don't let it stop you from getting your thesis written up for publication as quickly as possible. It is much easier to write your thesis up while it is still fresh in your mind than a year later when you haven't thought about it at all. One idea that some people advise is that when you write your thesis, you should make every effort to write as much of it as possible in an article form. This usually involves putting any material you would not include in an article into a separate chapter. Then when it comes time to write up your thesis, you write up a new introduction and cut out the material that is unneeded.

This approach has many advantages, but a few words of caution must be given. First of all, there are many mathematicians who feel quite strongly that this is not how it should be done. If the referee feels this way, he may subtly hold this against your paper. Also, be sure that you have taken out all of the inappropriate material from your thesis. A second disadvantage is that a thesis is different from a paper, and occassionally these differences are hard to edit out. What can happen is that your final paper reads badly because of the editing job. On the other hand, there are several advantages to this approach. Since your advisor has (presumably) read through your thesis, the odds are better that the mathematics is correct. If you rewrite your arguments or try to improve on them, there is a chance that you will add false statements to the paper. Of course, you also save some time and trouble by not having to rewrite your thesis all together.

One of the most important parts of any paper is the introduction. Most mathematicians will read the introduction of a paper to see what the paper proves, to see whether it is of any use to them, and to see whether the result or techniques are interesting enough to merit reading through the whole paper. The introduction of the paper is also what an editor is likely to read when deciding who should referee it. If your introduction does not make clear what the paper is about, you may find that the editor will choose a referee that is inappropriate. I state this last from painful experience. I once sent out a paper with a very bad introduction and paid for it. The first referee at the first journal panned the paper. Later after rewriting the introduction and making other minimal changes, the paper was accepted by another (more prestigious) journal.

The above make it pretty clear what the introduction needs to do. Your main results should be clearly stated in the introduction. This doesn't mean you have to state them in fullest generality. Rather, if a slightly stronger than necessary hypothesis makes the theorem a whole lot easier to state, you can state it that way in the introduction. In this case, however, you should mention that you actually prove a stronger result. The introduction should also put the paper in some kind of perspective for the reader. Mention similar results others have obtained, and if your paper answers a problem raised elsewhere, be sure to mention this. The point is that you need to motivate your paper. One hopes the referee will know the motivation behind the result, but it doesn't always happen. Journals want to publish papers that are of interest. Without motivation, the interest of your paper is unclear. For example, suppose you are writing a paper on solving a particular equation in number theory, but the reason you want to solve it comes from finite group theory. It is possible your paper will be refereed by a number theorist who will find the result boring. Thus, you need to explain the interest to group theorists. At least in this case, the editor will know who to contact if the referee questions the motivation.

Another important aspect of paper writing is knowing your audience and writing to them. By this I mean that your paper should be different if you are writing for a general audience instead of for complex analysts. An *MAA Monthly* paper should be written differently from a paper submitted to *Math Annalen*. A paper in the *College Mathematics Journal* should probably be written differently from a paper in the *Monthly*. Also, if you are writing a paper in analysis that has applications in number theory, try and write it so a number theorist can read it. There are two reasons for this. The obvious one is that you want your results used. If no number theorist can understand your paper, you make it harder for them to use it. The second reason is that your paper might get refereed by a number theorist, and if the referee has trouble reading it, life is a lot more difficult.

After you have written the paper, you should show it to colleagues (and your mentor) if you have one. They usually will have some very good suggestions as to how to improve it, and will find the places where you should add some detail. You should also make sure doubly sure of the mathematics. The refereeing process is not meant to take the place of careful revision by the author. Ultimately, you are judged on the results in the paper, and you take the heat for an incorrect proof. I have found it helpful to put a paper aside for two to four weeks after writing it, and then go back and edit it again. This way, I find out what was really obvious and what wasn't. Usually, when writing a paper, I am so immersed in the subject that I don't even realize that something that is obvious to me is not obvious to others. When I go back to the paper, I discover these problems. On occassion, I even discover errors that these jumps cause. Most of the time, one month doesn't mean much to a paper in the grand scheme of things. What you want to avoid, however, is having a paper in the revising stage for a year. Hence, at some point you have to send off the paper even if you haven't yet gotten the very best result you could hope for. I don't mean that you should send off inconsequential results, but rather, don't hold off publishing that major result just because you might be able to weaken the hypothesis still further in another six months. Paul Halmos has written a very good article on communicating mathematics—it's witty, learned, and highly recommended (see [3] below).

As far as disseminating your work, you should probably create a preprint list. This is a list of mathematicians who are interested in your field. The idea is that you send preprints of your papers to these mathematicians. This way they know what you have done and what you are working on. It provides one way to try to prevent duplication of mathematics. Also, this is a very good way to find out about other ideas. I once sent a preprint out and received a reply asking how these techniques might work on a problem in another paper. Since I didn't know the other paper existed, I hadn't known about this problem. It turned out that the techniques were exactly what I needed to solve this problem. While I didn't get another paper out of this, I did improve the paper I had.

The last comment to make is that you need to write up your results, but you also need to continue doing research. If you take a year to write up your thesis and haven't been doing research at the same time, then you could be in big trouble come tenure time (at universities requiring research for tenure). Keep working on improvements on your thesis, and also try to branch out. If you are lucky, your thesis will lead to a bunch of other papers, but if you are unlucky, your thesis might lead up a blind alley. In this case, it is important to have other avenues of research started.

There are many resources to consult about mathematical writing. See the references at the end of this chapter for examples.

Where to Publish
by Charles Holland

In selecting a journal for submission of a paper, there are several considerations to bear in mind. The paper should have some chance of appearing in a timely fashion. It should be in a journal with wide distribution. It should be a journal with a decent reputation. It should be a journal where the paper is likely to be refereed by a knowledgeable and sympathetic referee. Finding the necessary information will probably require spending some time in a library with a good selection of journals. Of course, once the decision is made, you need to look at the journal to see specifically what they require in the way of format, number of copies, and to what address it should be sent.

Twice each year, the *AMS Notices* publishes the backlog and estimated waiting times for papers submitted to many of the standard journals. Other things being equal, I try to avoid journals that are too far behind. Next, if my paper is closely connected with other papers, I consider the journals where those papers appeared (or where their referenced papers appeared), on the grounds that it will be easy for the editor to find a good referee. For similar reasons, I check the list of editors of the journal and try to find someone who I think will be interested. Of course, some journals have specific restrictions on subject matter, length, *etc.* There is a perceived ranking of journals by "prestige", though personally, I tend to discount that. If that makes any difference to the author or the author's university—our Dean once sent the department a list of "the twenty most important Math journals" with the suggestion that we publish more in them—then the only way to find out about the relative ranking of journals is by asking (and you will probably get different opinions from different people). In any event, it doesn't hurt to ask colleagues and friends for suggestions. There is a lot of informal information floating around which would not be obvious in any library search. For instance, certain editors are known to be "black holes"—the submitted papers disappear and are never seen again, and some journals have a definite tilt toward or away from certain subjects, though they don't necessarily publicize this. Finally, if the author believes the paper would be of interest to a certain mathematician, and no other obvious choice presents itself, then send the paper to that person and ask for an opinion on publishing options.

Where to Publish
by Steven Krantz

You've just written a paper. You think it's pretty good. Where should you submit it?

It is difficult to judge the genuine worth of one's own work, especially when one is still in the heat of passion. Before the problem is solved, it seems mightily important. After it is solved, one is tortured by self-doubt: shouldn't I have solved this much more quickly? Is it really all that interesting? Could I possibly get it published?

One good barometer of how to proceed is to show your preprint to friends and colleagues. Are they surprised, impressed, confused, bored? Sometimes they will suggest changes. Sometimes they will suggest a journal. (Sometimes the friend is the editor of a journal and will offer to communicate it, but don't bet on this—it is obviously the exception.) Ultimately, the decision of where to submit is up to you.

Begin by considering where cognate results have appeared. The *Journal of Algebra* will probably not consider papers on singular integrals. The *Journal of Symbolic Logic* probably doesn't publish papers on Gelfand-Fuks cohomology. Certain journals have become the default forum for work on operator theory or several complex variables or potential theory. It is natural to consider those. It is also natural to consider which editors will understand what your paper is about and will know how to select a referee. You need not actually *know* the editor, but it is comforting to know where the editor is coming from.

If you shoot high and select a really classy journal, you might pay in several ways: 1) the refereeing process may take an extra long time, 2) the journal might have a huge backlog, 3) the paper could get rejected for almost any reason. Thus, the entire process of getting your work published could drag on for two years or more. If you are fighting the tenure clock, this could be a problem. In some ways it is better to err on the side of shooting low. Usually mathematical work is judged on its own merits. Nobody will downgrade your work, or you, if your theorems are not published in the optimal journal. But don't publish in an obscure journal that nobody ever reads. There are some journals with the reputation that they would publish Dan Quayle's laundry list; that is not where you want your work to appear.

There is no magic formula for picking the right journal to which to submit your work. In some ways it is a crapshoot. If the paper is rejected, it is not always correct to conclude that the paper is worthless. It may have landed in the hands of a referee with an axe to grind, or who did not understand it. The editor might have misread the referee's report. Don't be afraid to get help or advice on how to proceed.

Some of the best mathematicians that I know—even Fields Medalists—can reel off horror stories of all the papers that they have had rejected. Part of surviving in this profession is learning to live with the reality that one needs to persevere. If your paper is accepted first time around, then congratulations. If not, try to be objective and figure out why. Then act intelligently on that new information.

References

1. R. Boas, *Can We Make Mathematics Intelligible?*, American Mathematical Monthly **88** (1981), no. 10, 727–731.
2. L. Gillman, *Writing Mathematics Well: A Manual for Authors*, MAA, Washington, D.C., 1987.
3. P. Halmos, *What To Publish*, American Mathematical Monthly **82** (1975), no. 1, 14–17.
4. N. Higham, *Handbook of Writing for the Mathematical Sciences*, SIAM, Philadelphia, PA, 1993.
5. D. Knuth, T. Larrabee, and P. Roberts, *Mathematical Writing*, MAA Notes Series, vol. 14, MAA, Washington, D.C., 1989.
6. S. Krantz, *A Primer of Mathemtical Writing*, AMS, Providence, RI, 1997.
7. N. Steenrod, P. Halmos, M. Schiffer, and J. Dieudonné, *How to Write Mathematics*, AMS, Providence, RI, 1973.
8. *A Manual for Authors of Mathematical Papers*, AMS, Providence, RI, 1990.

CHAPTER 6

Getting Grants

This chapter offers the advice and experience of a very diverse group (diverse in the grant proposal sense, at least). One author has never gotten a grant; others have recieved one or two; still others have been awarded many. Some of the authors are in the trenches of proposal-writing; other authors read the proposals. But all of these authors repeat the same advice:

(1) Apply for grants, and when you get rejected, apply for grants again.

(2) Talk with a real person from the granting agency.

We begin the chapter with advice from mathematicians who ask for grants, and wind up with advice from the people who distribute them.

What I Learned Applying for an NSF Grant
by Mary Shepherd

My name is Mary Shepherd. I graduated from Washington University in St. Louis in May, 1996. I was a nontraditional graduate student who returned to graduate school after 15 years in private industry most of the time working as an accountant. I returned to school because I wanted to teach, not because of a great love of doing research. In fact, I was very nervous about doing original research when I returned to graduate school. To my surprise, I found that I did enjoy the research, although I do feel I will not be one of the world's great researchers. I am starting my second year at SUNY-Potsdam in a tenure track position. Potsdam is a public liberal arts school with about 4300 students. My teaching load is 12 hours per semester. Our department has 10 full time faculty.

Last year in September all the new faculty here were invited to a retreat where we were given an orientation to the school. Among the many presentations was one by the Faculty Scholarship and Grants (FS&G) office. I let most of what they said go in one ear and out the other. I was on information overload, but I did become aware that our school has an office to help faculty write grant proposals and administer them after a grant is received. Being overwhelmed by the teaching load, though, I gave it little other thought. I found I was struggling to stay even with the work in the classes I was teaching.

In January, during our winter break, I managed to write a paper out of my dissertation and send it off to a journal, but new research did not happen. Around late March, all faculty received a newsletter from our FS&G office with some grant programs listed in it. I am sure we had been receiving these frequently all school year, but I had usually managed to ignore them. I figured I was not the grant type person. After all, only really great research ideas are funded by the NSF, aren't they? And I didn't know any other funding sources for mathematical research. With the newsletter there was a note that anyone who might be thinking (ever) of applying for a grant should set up an appointment with the director of the FS&G office. Since I thought it was possible that someday in the far distant future, I might apply for a grant, I went in to talk to the director of the FS&G office over spring break. We talked about how people just out of graduate school go about getting their research disseminated—that first article, eventually invitations to speak on it, *etc.* I found that our FS&G office helps fund travel when we give a talk, and other miscellaneous information. I still did not expect to be applying for a grant in the near future.

After our meeting, though, I went and read the newsletter. I found a program that seemed to match me. It was an NSF program, so I went to the NSF web page `www.nsf.gov`and got more information. I discovered that the match was not as good as I thought, but I found another program which seemed to match better. When the spring semester ended, I decided to try and apply for that grant. The deadline was July 22, so I figured I should be able to do it. I had an idea for research, discussed it with my advisor, and he said "go for it". So I did. In the end, for that proposal, I made a small mistake and the proposal was rejected, but I will be trying again in the fall for another program with a similar proposal.

In the process of applying for a grant, I learned these lessons. I hope they will help other young mathematicians.

1. Check to see if your school has an office such as our FS&G office to help you. They did the picky detail work on the grant application, the copying at the end and making sure everything was put together in the correct order with all signatures on the proper lines.
2. Decide to apply for some grant program. Try it—the worst anyone can say is no, and the no answer usually comes with comments that will make your next proposal better. If you don't do it at all, then the answer is automatically no.
3. Check out different sources. The NSF web page is a good place to start.
4. Get the instructions and follow them to the letter. If you have a comparable office to our FS&G office, they might help you with this.
5. Talk your proposal over with your department chair. You will need his/her support, also.
6. Allow yourself plenty of lead time. Check your deadlines and make sure to meet them. It will take at least 6 weeks of hard work to prepare a good proposal. Summer or winter break might be good times to start the process.
7. Call the contact person for whatever program you choose and ask questions. This can be a big source of help in both the design and presentation of the proposal.
8. Your advisor probably has some ideas to incorporate into a proposal. Mine had some good suggestions on timing of certain aspects and some equipment ideas, also.
9. Get lots of feedback on your proposal. Ask others to read and comment on your proposal. Use people in your department (those with grant experience and those without), people in any support offices such as our FS&G, your

advisor, and any collaborators if allowed. I revised my first proposal about 10 times.

10. If you are turned down, seek other sources and try again. Review any comments made and incorporate them into your next proposal.

11. Plan to have your proposal complete at least one week before the deadline. This will give you a little leeway if a correction needs to be made, or computers glitch or ...

In summary, why do it? In my case, it helped focus what direction I wanted my research and scholarship to take. It forced me to make plans and write them down. It made me want to get back into my research again. I doubt I will ever be the great researcher many of you can be, but I found that the biggest benefit to writing a grant proposal is that I am once again excited about trying to do some research and writing about it. So my advice to young mathematicians is to look around for grant opportunities and commit yourself. In the words of a shoe company, *Just Do It!*

When at First You Don't Succeed ...
by Curtis D. Bennett

Young mathematicians interested in research positions should apply for a National Science Foundation (NSF) Postdoctoral Research Fellowships. (In fall of 1998, the NSF began requiring that applicants be no more than two years from receiving their Ph.D.s.) Personally, I applied twice. The first time was my last year of graduate school (the fellowship would have covered my first three years out), and the second time was 15 months after receiving my Ph.D. The first proposal was unsuccessful, but on the second try I received a fellowship. Even today, I never expect for a proposal to be funded on the first try. Below I will discuss the differences between the two proposals and why I think the second one was more successful than the first.

In my first application I didn't have a clear problem to focus on. I basically said, "I want to continue the research topic of my thesis." In my second application, I gave a detailed background of the research problem I was trying to solve, included all of the partial results I had at that date, and tried to explain why the problem was worth solving. All of these made the proposal more focused. They also gave the reviewers reasons to believe that I could be successful at completing the project, and it gave reviewers that didn't know much about my research area reasons for supporting the project.

My choice of sponsoring scientist was better in the second proposal. In the first proposal, I listed one of the top people in my area as sponsoring scientist, but he was unfamiliar with my research. For the second proposal, I listed someone who was much more familiar with my research. I believe a strong letter from your sponsoring scientist is essential.

The other major difference between the two proposals was the letters in support of my application. The second time around I was fortunate to get a letter from one of the very best mathematicians in my field. Additionally, for the second application everyone who wrote a letter for me knew my research very well. As is often the case when you are graduating, I had one or two professors recommend me, but they presumably could say little more than, "I had Mr. Bennett in class Y and he was good."

In general, I think you should plan on submitting your grant more than once. That said, it is probably pointless to submit exactly the same proposal two years running. Rather, based on whatever feedback you can get, rewrite the proposal to correct errors in the first proposal. Just like mathematics papers, a well-written proposal is more likely to be funded.

You should also consider applying for a NSF (or National Security Agency (NSA)) summer grant. One drawback to the NSF postdoctoral fellowship process (at least as of 1993) is that you do not get feedback from reviewers. In my own case, the feedback I received from a summer grant application gave me valuable information for my second postdoctoral proposal, and I believe this information was crucial to my success. Finally, you should take a look at successful proposals from other years. The YMN archive contains four such proposals. The NSF website at www.nsf.gov also has a searchable index of proposals, although in my experience you can only get abstracts of past proposals, not the full proposal text.

If you are interested in a research career you should apply for an NSF postdoctoral fellowship. Don't expect success on the first try, but do give it your best shot, and then use the information from the first try to give yourself the best chance you can later.

Apply for an NSF Grant
by Daniel Lieman

NSF grants are hard to get, which makes some people reluctant to apply. It does not help that some people have large grants year after year, and are quite visible, while most get nothing. But the fact is that NSF grants do not all go to people at Princeton/Harvard/Chicago/Berkeley, and that they do not all go to senior established researchers.

All NSF information is publicly available. You can visit the NSF Division of Mathematical Sciences web site at www.fastlane.nsf.gov. You can find out every NSF grant given out in the past 7 years. All summaries are fully keyworded, so you can see who's getting the money in your field, after a quick search.

Probably the best advice to someone applying for NSF grants is to: a) Apply early—as early as is possible! and b) talk to the program director before you apply. These people know a lot, and will tell you if you ask. They can tell you what they look for in a grant, and against what criteria your proposal will be evaluated.

Grant Writing Basics
by Tina Straley

This article is a short summary of presentations that I have made on grant writing, in particular, at an MAA-SE Section version of Project NExT. In order to compress my remarks, I have written this in outline format. You might want to take this list and talk with someone on your campus who has grant experience and cover some of these points.

Where to look for grants.

1. Federal agencies: the Department of Energy, the Department of Education (FIPSE and K-12 curriculum programs), NSF, NIH, NASA, NSA. (Other than the Department of Education, all support mathematics research and curriculum development)
2. Private foundations: Exxon, IBM, *etc.*
3. State agencies.

How to structure the proposal. Although each of these agencies and each program will have somewhat different solicitations and formats for proposals, there are enough commonalities that there are some general guidelines that pertain to all proposals. However, always follow the format and requirements of the grant solicitation; do not leave out sections, and do not simply write for one program and repackage for another without revising and responding to the call.

Some general guidelines. Stay within page limits; don't use super small fonts or cut down on margins to get more in. It is clarity and substance that counts, not quantity. Attach appendices if allowed, but do not make them a continuation of the narrative—these should be supplementary materials. Check grammar and spelling as the last thing you do. Be clear and concise. Get someone else not connected with the project to read a draft and comment. Take advantage of submitting pre-proposals if allowed. Talk to a grants officer about your ideas during the development stage.

The factors that make the difference. The idea must be innovative and something the reader will feel "I wish I had thought of that" or "I wish that I could do that." It also must fit the program. Show departmental and institutional support. Show that you have the right individuals or teams to carry out the project, that you are qualified. Present a reasonable plan to implement the project or carry out the work. Show that you are knowledgeable in the field (education or research), that you know what has been done along similar lines and how your idea is different or an improvement. The idea should not be pie-in-the-sky. A proposal is much stronger if you have tried out the idea and can show it works, or if the idea builds on previous work you have already done.

And finally .... Here are some suggestions for getting in the grant writing business. Work with a seasoned faculty person in order to establish your own reputation and credibility. Get someone to review your proposal for you. Read the announcements of awards so that you can determine if your ideas fit the thrust of the program and you are not reinventing the wheel; you will need to establish how your project adds to the field of knowledge and how it makes a difference. Include a bibliography and refer to other projects/results within the body of your proposal. Don't give up or become discouraged if at first you do not get funding. Many well known people have written more unfunded proposals than funded ones. In fact, if you know of someone who has lots of grant money you can bet they had proposals rejected as well. Read reviewers' comments carefully, don't reject out of hand that they did not understand you. Consult the program officer if you don't understand the reviews.

Finally, always look at proposal writing as a growth experience in itself.

A Perspective on the Division of Mathematical Sciences at NSF
by Robert Molzon, former Program Director at NSF

Introduction. I would like to try to provide some information about the NSF that might be useful to recent Ph.D.'s in the mathematical sciences. I would also like to provide my perspective on the direction NSF might take in the near future in order to address some of the difficulties faced by the mathematical community and recent Ph.D.'s in particular.

I should first mention the article written by two former Program Directors, B. E. Trumbo and Russ Walker which appeared in the September 1990 issue of the Notices of the American Mathematical Society (vol. 37, pp. 838–843). Although written a few years ago, the information is still quite relevant and should provide a good guide to preparing a proposal. Before writing a proposal for submission to the NSF I certainly suggest reading their article in order to obtain a basic understanding of the award process.

In this article, I shall first present a brief outline of the organization of the Division of Mathematical Sciences and of some ways in which mathematicians might improve communication with the NSF. The importance of getting to know Program Directors is stressed in the article by Trumbo and Walker, and I endorse what they say. I also suggest a new approach to support of research that could be taken by the community. Although the approach I present is not completely new since several proposals of the type I mention below have been funded, I believe the approach will be new to most people.

Organization. The Division of Mathematical Sciences (DMS) is one of the divisions in the Mathematical and Physical Sciences Directorate at NSF. The DMS is broken up by area of research into ten programs. The major disciplinary programs such as Geometric Analysis are administered by one or two Program Directors who have primary responsibility for recommending awards within the program. The Special Projects Program is primarily responsible for proposals for large group projects such as institutes and conferences. In addition to the disciplinary programs, DMS participates in a number of initiatives which support researchers in fields related to the initiative. The High Performance Computing Initiative is an example. Funds earmarked for the initiatives can be transferred into a Program in order to support work on a project which has a component closely related to the initiative. How directly related a project must be to an initiative in order to qualify for support by initiative funds was never very clear while I was a Program Director. If one is working in an area which might be related to one of the current initiatives, it would certainly be worthwhile contacting a Program Director for guidance in writing a proposal so that the link to the initiative is clear.

Approximately one half of the Program Directors are "rotators" and one half are permanent NSF staff. The "rotators" serve one or two years and continue to hold their university appointments during their tenure at NSF. It has generally been very difficult for NSF to entice mathematicians into serving at the NSF for one or two years—a problem that seems to be unique to mathematics as other sciences do not have this difficulty. Program Directors do have a fair degree of autonomy within the constraints of general Division and NSF policy. That autonomy is limited to the individual programs which they direct and of course is constrained by budgets and the established format of the NSF award system. Trying to make substantive changes in the system proved to be a daunting task—even in cases when a large majority of the Program Directors felt change was necessary.

Communication with the DMS. A large percentage of the Program Directors' time is spent communicating with the mathematical community. Much of this is routine and involves passing on information and answering queries about the status of proposals. As Trumbo and Walker point out in their article, it can be very helpful to contact a Program Director at NSF for assistance in preparing a proposal. They can provide first hand information which may not be contained in the standard publications of NSF and might not be generally available to grants officers at a university. It also provides an opportunity to talk in an informal way with the Program Director and simply establish some personal contact. Your department can also invite a Program Director to give a colloquium. This allows a large number of people at your university a chance to get to know the person making the decisions on research proposals. Surprisingly few departments make this a regular practice. I believe the ones that do have really benefited—both in terms of obtaining information on proposal writing and in getting their concerns across to the Program Director. In general, the Program Directors have a great deal of sympathy for the problems faced by mathematicians with respect to grants. It really does help them to know what your problems are and it especially helps to know that they have your support when they are trying to make changes within a large bureaucracy. The personal contact that comes from a visit to a campus is beneficial to the Program Director as well as the Department, and they will make every effort to make the trip. Because of the heavy work load associated with writing recommendations on proposals in the Spring, late Summer or Fall is the best time for an invitation. The cost of these visits must (by NSF rules) be assumed by the government; hence, even departments with zero colloquium budget can afford it.

Nontraditional Proposals. Almost all of the proposals submitted to the disciplinary programs follow a single format in their request for funds. The traditional proposal is submitted on behalf of one or two principal investigators and the request for funds follows the pattern of two months of summer salary support, travel funds, and perhaps funds for miscellaneous items such as page charges and telephone calls. A number of proposals I handled as Program Director had six or seven principal investigators, but the breakup of the budget request followed the traditional format. With the implementation of some form of "flat rate" funding this format may have changed slightly, but the most significant direct cost remains summer salary.

With increased pressure on program funds due to relatively small or no increase in program budgets, more people have begun to question the traditional NSF award in mathematics. A recent Forum article (vol. 41, February 1994) in the Notices of the AMS, "Some thoughts on the funding of mathematics", by William Yslas Velez, former Program Director of the Algebra and Number Theory Program, summarizes some of the different points of view that have been taken by various factions in the debate.

Virtually all of the proposals submitted to the disciplinary programs, with the exception of those for support of conferences, involve research which is carried out over twelve months, not two summer months, and hence the common assertion that a traditional NSF award in mathematics is meant to fully fund a research project is certainly questionable. Keeping this in mind, it might be useful to ask what type of support is most necessary to carry out a research project. Since traditional NSF awards fund at best only a very small percentage of any research project, perhaps emphasis on the most necessary items will allow mathematicians to write proposals which are attractive to NSF because they set priorities and present cost effective alternatives to traditional proposals.

Small travel grants have frequently been proposed as a means of providing some much needed support to individuals who have strong active research programs, and whose most pressing need is travel to work with colleagues or attend conferences. Because of the administrative burden of processing proposals and awards, this is not a cost effective way of supporting research. This should not mean however that mathematicians cannot set priorities in seeking research support; it simply means that proposals must present a cost effective approach to NSF.

One point which seems to have been missed almost entirely by the mathematical community is that the NSF does not require that proposals follow the traditional funding pattern—in fact NSF guidelines are quite flexible in setting proposal format. The most important point I would like to make in this article is that one is not locked into the traditional approach.

As an alternative to the traditional approach to funding I would suggest that several individuals working in closely related fields at a particular university write a joint proposal that details how funds for travel, colloquium speakers, and student support would help their research program. For example, four or five mathematicians in a department working in the general area of differential geometry might be holding joint seminars and be interested in inviting visitors to speak in a colloquium. They might also be teaching related courses which are attended by graduate students interested in geometry. I believe such a group which had a strong research program could write a very attractive proposal for support of visitors, students, and travel. A joint proposal by a group *should not* look like four or five individual proposals stapled together. Rather the proposal should emphasize the group activities, and how support of the group will benefit a research program. Of course individual research accomplishments and goals should be described, but the benefit of having strong group interaction should be emphasized.

When I encouraged group proposals, I was asked if the individuals must all be in one field of mathematics or writing joint papers. I believe the proposal would be stronger if the individuals could show common interests and good communication between members of the group. This certainly does not mean the individuals must all be in one field of mathematics. In fact, if one could show common interests among individuals in diverse areas this might well be viewed as a strong point— especially in light of the current emphasis on interdisciplinary work.

One potential difficulty of group proposals of this type is the possibility of flooding DMS Programs with proposals. This would place a very serious burden on a staff that already has the highest proposal load in the NSF. I believe it would be much easier to convince a Program Director that one has a serious need for support by submitting one well written and thought out proposal than by sending in multiple proposals. The groups for which I recommended funding were quite conscientious about this point.

Finally, I would emphasize once more the importance of contacting the appropriate Program Director for information if you believe you could write a strong group proposal. Different Program Directors may well have different ideas on points to stress for proposals of this type.

Summary. With the increased difficulty of obtaining even modest support for outstanding research projects in mathematics, many in the mathematical community may have adopted a rather resigned attitude with regard to NSF support. In this short note, I hope I have outlined a possible alternative to the standard manner of seeking NSF support. If this approach is to be successful, it will require good communication between the mathematical community and the Division of Mathematical Sciences at NSF. The community can take a step in the communication process by getting to know the Program Directors and by writing to the administration at NSF.

I would like to thank the Young Mathematicians' Network for the opportunity to present my ideas on a direction that the NSF and the mathematical community should seriously consider in the search for a way to address the current crises in research support for mathematics.

CHAPTER 7

Tenure

You can't get tenure all by yourself.

One of the most important things you can do to prepare yourself for the tenure process is to build a network of friends and mentors—colleagues just a few years ahead of you, senior faculty in your own department, senior faculty in other departments, and mathematicians from outside the institution—who can offer you opinions and frank advice about your institution and your own performance within it.

Applying for tenure is fraught with questions, and the answers are usually complicated and subtle. Actively building such a network of supportive colleagues can help you understand the answers to these important questions:

What are the particulars of your institution's tenure review system? Is tenure decided by a committee or by an individual? What are the weights assigned to each aspect of your job? How many letters of recommendation will you need (from students, people inside your department, outside your department)?

How does your institution measure success? Do talks at Special Sessions count for research? Is teaching measured by in-class criteria alone, or does guiding Independent Studies or Putnam teams matter, too? How should you interpret your student evaluations?

How are the political and economic conditions at your college likely to affect tenure decisions?

Since judgments about teaching are frequently more subjective than those about research, other faculty can help you with a before-it-counts reality check. You will learn a lot if you have another instructor sit in on your course, and also if you visit courses taught by other successful professors. Have faculty members provide *specific* comments on teaching, and get a mentor to work on teaching with you.

Especially at places where politics may play a role in the tenure process, it helps to get a senior faculty to act as your advocate.

Pay attention to how job applicants are evaluated during times of hiring—this should tell you something of how you will be evaluated during the tenure process. Make sure you understand the tenure process in advance.

Of course, there are parts of the process that are within your own control. The tenure decision is based on what the administration sees or knows about you, so you should actively point out your accomplishments. "Toot your own horn," a guidance counselor once told us. You can and should conduct your own assessment means: collect student evaluations, keep tabs on how many of your students take more math classes; count your citations in *Math Reviews* and *Science Citation Index*. Document your teaching effectiveness at elementary as well as advanced levels.

The bulk of this chapter is Dana Mackenzie's excellent (and harrowing) description of his own experiences with the tenure process. To lead into this, we begin with a brief essay on the "murky business" of evaluating teaching.

Interdepartmental Evaluation of Teaching
by Leonard VanWyk

Evaluating teaching performance is a subjective, murky business. The two most common techniques used are student evaluations and some form of mentor review, usually by the Chair of the department. Both of these evaluation methods glean certain information about the instructor, but both have their weaknesses as well.

While student evaluations can give valuable information about the students' perspective, they can easily degenerate into superficial popularity contests. On the other hand, an evaluation from the Chair of the department can provide insightful suggestions from a more experienced instructor, but the Chair's judgement is inherently biased due to his/her own experience in teaching the same material. Thus each of these two methods yields certain information, and together these two perspectives paint a rough, limited picture of the instructor's teaching ability and style.

In order to sharpen this picture somewhat, why not introduce the concept of an external review by a faculty member outside of the department? This person, who could be chosen from any field for the evaluation of a lower-level course, would sit in for a class or two, similar to the Chair of the department. This type of evaluation provides the insight of an individual who knows something about teaching techniques in general, but little or nothing about the particular material; in some respects, this would be the most informative of the three types of evaluation discussed here.

This third type of review is not without its problems. It would require co-operation between departments—a BIG problem at some schools—as well as a willingness for instructors to accept suggestions from, *gulp*, outsiders. However, it would be relatively simple to implement and quantify compared to, say, measuring the success rate of the instructor's students in subsequent related classes.

The fact remains that teaching ability is not accurately measured, even though evaluating this ability with the greatest possible degree of precision should be a priority for many institutions, given the weight that it can carry in the tenure decision.

The Tenure Chase Papers
by Dana Mackenzie

Foreword. In recent years, the mathematical community has become concerned about the talent drain that occurs when new Ph.D.'s cannot find academic jobs that match their expertise, or cannot find jobs at all. Edward Aboufadel's "Job Search Diary," published in *FOCUS* a few years ago (1992–1993), gave a very insightful account of a new Ph.D. entering the uncertain job market for the first time. However, I have never seen an article that specifically addresses what to expect at the next great hurdle in an academic career, and another point at which a talent drain undoubtedly occurs: the tenure decision. In this series, I would like to remove some of the veil of silence and mystery that surrounds this process, by chronicling my own experience as I sought tenure at Kenyon College, a small liberal-arts college in central Ohio. Although there are many points in the story that are unique to my case, I am making this chronicle public only because I believe that my experience has some valuable lessons for anyone who may be connected with a tenure decision. I hope that junior faculty will learn ways to improve their chances and warning signs to heed seriously. For tenured faculty members, I hope that there will be some lessons on the need to mentor junior faculty and to intervene on their behalf when circumstances require it. Finally, for the mathematical community at large, I hope to provoke some debate about whether the institution of tenure is working in the way that it should, and whether we might be better off without it. In the narrative below, the parts written in italics were written specifically for this article, after all the events recounted had occurred. The parts written in normal text are excerpted from my personal journal, and were written at the time the events occurred. I have removed personal identification wherever possible and have abridged many of the entries, but otherwise the content of the entries has not been altered.

I would like to acknowledge the advice and support of my wife, Kay, not only in preparing this article but also throughout the tenure chase. Without her, this story would have been much less interesting because I would have given up too soon!

Contents. 1. Prehistory 2. The Ax Falls 3. Grievance 4. Double Jeopardy

Prehistory. *After earning my Ph.D. from Princeton University in 1983, I taught for six years at Duke University. I left when it became clear that I was not going to get tenure. At that time, the Duke math department attached a great deal of importance to research accomplishments. Although I had been awarded a grant from the National Science Foundation in 1987, this was barely enough to "keep up with the Joneses" in a department where, remarkably, every tenure-track faculty member had outside funding in 1988. While at Duke, I gradually grew more interested in the human interactions of teaching, and my enthusiasm for the more solitary work of research lessened somewhat. But there were no rewards at Duke for a commitment to teaching. When I arrived at Kenyon in 1989, I was pleased to find that the quality of teaching was taken vastly more seriously there.*

In rereading my diary from my early years at Kenyon, two things stand out: how little I actually worried about getting tenure, and how much I still judged my success on the basis of research, rather than teaching. I didn't worry about the tenure decision because Kenyon had a very high tenure rate in recent years (according to many people, the second re-appointment was the most critical review, and the tenure rate was 100% for faculty who got past that point), and because the mathematics department was clearly very satisfied with me. My attitude towards the relative importance of research and teaching was colored by the prevailing view of the mathematical community, that "success" equated to proving theorems. I was soon to find that this was not the prevailing view at Kenyon. The first wake-up call came when I had my second re-appointment review in 1992.

5/11/92: On Saturday I received my evaluation from the provost, on which my re-appointment and merit raise were based. It was not what you would call glowing, particularly with regards to my teaching, which he described as "uneven." There were some good points made and some specious ones ... provost observed that "the better the student, the better the evaluation." My first reaction was "Duh ... What else is new?" That's the way it always was and always will be. But then he made a comment that got me thinking: "That suggests that you need to reach out more skillfully to the weaker students." ... I am still very impatient with students who don't make an effort. My philosophy tends to be (to put it charitably), "They're grown up, I'm not their baby-sitter; if they choose not to work hard, they can live with the consequences." I freeze them out, rather than talking with them. To the extent that I do try to make contact with them, it's definitely not "skillful." After the [most recent] Math 11 exam , I let the next class out early and said that I would like to talk with each of the people who got below a C. I talked with half of them (the other half weren't in class), and the conversations I had with those three were awkward, embarrassing, and in two out of three cases, quite unproductive.

Even as I write this, though, I wrestle in my mind with the question of how much of a change I can really make. I shouldn't have to be a therapist or a guidance counselor; there are other people at the college who are paid for their ability in those realms. How "skillful" can I reasonably be expected to be in dealing with students who are at the low end of the motivation or maturity scale? I don't know; but evidently the answer is, more skillful than I am at present.

As you can see, the provost's letter raised anew questions about my own competence as a teacher that I had more or less resolved several years ago at Duke. Not that I had stopped being aware of the difficulty of the job or the fact that my personality is in some ways unsuited to it, but basically I have evolved a modus operandi which allowed me to feel that I had improved and that I had learned to do the job fairly well. I would say that my confidence is now shaken. I should, perhaps, admit that one reason for the strength of my reaction was that the letter also presented (as is apparently required by the faculty bylaws) "grades" on my performance in three categories: Teaching Excellence: B-; Scholarly Engagement: B; Collegiate Citizenship: B-. Back in my student days I never received grades that low on anything, so, as Kay correctly observed, my pride was wounded.

6/18/92: [I met with] the provost to discuss the findings of my second re-appointment review. In general, [the meeting] was positive and supportive. One of the main pieces of information I wanted was how to interpret the "grades" that he gave me. He made two relevant comments. First, everyone who was reviewed was "graded" on the same scale, whether they were up for second re-appointment or promotion to full professor. It was no surprise, then, that the grades for those in the former group were somewhat lower than the grades for the latter group. For the college as a whole, the provost said, the median should be considered to be around B; for those in the second re-appointment cohort the median would be lower.

Gradually, I recovered from the shock of the "grades," and later events rein-forced my impression that they were simply an aberration. There were enough com-plaints from other faculty members about the grading system, which had just gone into effect that year, that the experiment with grades was abandoned after 1992. Moreover, I received the following news a year later that made me feel as if getting tenure would be a cinch:

6/18/93: ... Good news came in the mail today. I am going to receive the George Polya Award from the Mathematical Association of America, given each year to the two best expository articles in the College Mathematics Journal ... This is the first real public recognition I've gotten for mathematics since I got my NSF grant in 1987, and I'd have to rank it with that as a highlight of my career so far. It's certainly the best thing that I could imagine happening to me now, with a tenure decision coming up next year ...

If the provost only saw fit to award a B to my "Scholarly Engagement," when the MAA judged a part of it to be worthy of a prize, how seriously could I take his other comments? Unfortunately, I failed to grasp that the important thing in the tenure decision would not be reality but the administration's perception of reality.

Meanwhile, I continued to work on the real and imaginary deficiencies in my teaching that were found in the re-appointment review, but not always with success:

9/4/92: I was pretty dissatisfied with both of my calculus lectures this week. Both times I had to rush at the end of class, which was a specific problem I am trying to overcome this year ... I need to learn to pace myself and parcel out the time in a planned way. When I am running out of time I simply don't have enough control over events. For example, in Thursday's class I forgot to give the students a handout even though I brought it to class and wrote on my lecture notes: "DON'T FORGET HANDOUT!" Why? Because I was so rushed that I didn't look at my lecture notes in the last five or ten minutes.

4/8/93: Only six out of thirteen students came to my calculus class. When I mentioned that to Kay, she thought it was outrageous—both that the students would care so little, and that I would let them get away with it [by saying nothing]. So I did something about it. I sent the seven absentees a fairly stern reprimand by e-mail. But when I brought it home to show Kay, she said that ... I should have written it in a concerned, friendly way. Sometimes I feel that the harder I try to do the right thing, the less I succeed in doing it ...

One of the provost's comments when I was reviewed for re-appointment last spring was, "You need to reach out more effectively to the weaker students." ... This is the first time that comment has made sense to me.

I let students miss class because I hate confrontations. I don't like to do things that someone might consider "mean." I don't like to pry into other people's lives because they might think I'm "nosy." I don't like to ask favors because they might think I'm being "unreasonable." I don't like to insist on being listened to because I'm afraid that my audience just doesn't care!

But my students don't know these things. As far as they can tell, I don't care whether they come to class or not.

But not all my teaching experiences were so discouraging. My wife, Kay, wrote about the following incident in the Kenyon College Alumni Bulletin, August 1992:

> You've never lived until you've been awakened at ten minutes till
> seven in the morning by students phoning to say they want to

come over and bring your husband a great rhombicosidodecahedron. You've never lived, that is, in a Kenyon faculty household. I met them at the back door, ushering them in with silent gestures and pointing them toward their unsuspecting mathematics professor. They appeared at his elbow at the breakfast table, nearly causing him to choke on his English muffin. Michelangelo himself could have displayed no greater pride than that with which they presented the great rhombicosidodecahedron. Constructed from graph paper and gouts of glue, it resembled a giant, beveled golf ball. It was their favorite of all the Archimedean polyhedra ... "Close interaction between faculty members and students." I've heard it again and again, before we came here and during our years at the College. After the math students presented their treasure and departed, that phrase rose before me, suddenly gaining personal importance ...

11/1/92: Kay's article for the Kenyon College Alumni Bulletin ... has now been reprinted and is being mailed out to all the high school seniors that the Admissions Office contacts about applying to Kenyon! The reason is not for its mathematical interest, but because of the persuasive argument she makes in favor of small colleges ... The president [of the college] told us ... that the article had also been popular with the trustees, who kept Kay's boss [the editor of the alumni magazine] busy telling them how to pronounce that long word!

As the school year 1993-4 began, storm clouds gathered over the college, presaging a change in the economic and political climate in which my tenure decision was to be made.

9/30/93: ... [A biology professor] and I continued the discussion with [a history professor and a librarian] over lunch. They think that the president's reign has passed through three eras ... In the last few years [the third period], ... the president has lost hope of professional advancement and become more fiscally conservative, and the librarian said that she thinks he's even a little bored. The results of this are visible in the zero faculty growth, lack of support for grants, lack of real leadership on the science building, and the remarkable hysteria this year over the fairly modest shortfall in the number of students.

11/12/93: Yesterday the provost dropped the biggest bombshell from the administration that I have heard of since I came to Kenyon. Because of the financial hardship caused by the shortfall in enrollment, they (he and the president? or maybe just the president?) have decided to cut back on the number of faculty next year, in order to save money. This is going to be done by "suspending" several positions temporarily: not hiring replacements for faculty going on leave or retiring.

11/21/93: The decision was swift and unfavorable. On Thursday morning ... the chairman of the mathematics department [referred to henceforth as "the chair"] got the word from the provost that the math department will have to make do with five professors next year [instead of the normal six].

That fall also marked the publication of Alma Mater, a book by Kenyon alumnus P. F. Kluge, who returned to campus to teach and live for a year. A central theme of his book was what he called the "every kid a winner" syndrome, the gradual erosion of standards that leads to grade inflation. It also leads, in his opinion, to a situation in which an unacceptably high proportion of faculty members were receiving tenure.

Much later, a member of the College's senior staff told me, "I think that book really got under [the president]'s skin."

In spite of the warning signals that this might not be a good year to be coming up for tenure, I remained blissfully optimistic about my chances. In April, the hints became a good deal more direct, and my denial of reality shifted into overdrive. The next entry takes place just after the mathematics department finished undergoing a review by two external evaluators.

4/6/94: ... The chair reported to me separately a minor point that came up in the discussion [between him, the evaluators, the president and the provost] ... The evaluators reported on their meeting with the students on Monday night—which, incidentally, was very well attended, with about 25 students. There were glowing praises of three of the other professors in the department but after they were finished the president pointed out that they hadn't said anything about Mackenzie, and wondered if there was any reason for that. The chair said the question surprised one of the evaluators, who replied that there hadn't been any comments either positive or negative about me. I tend to put a fairly neutral construction on this observation and the president's question. There were no comments on me because the meeting was mostly for math majors and minors, and I just haven't taught very many of them this year (only two) ... The president asked because he know I am up for tenure and this was another good source of information. Of course, more insidious meanings can also be read into this exchange.

4/11/94: On Friday the chair had a mysterious meeting with the president and the provost. We figured it had something to do with the evaluation, but after the meeting he said it had not been about what he expected, and he was "sworn to secrecy." I have a wild guess. What would the administrators want to tell him about so urgently, so secretly, and so close to Honors Day? My hunch is that either Professor H or Professor S is going to win a Trustees' Teaching Award, and the administrators were letting him know so that he can make sure that they come to the ceremony.

4/21/94: My hunch about Honors Day turned out to be wrong. The winners of the Trustee Awards were ... Not Professor H or Professor S. Too bad. I don't know how they determine the winners of those awards, but clearly it's not by polling the math students.

4/24/94: The course of my life over the next several years has already been decided, but I do not know the decision yet. The trustees of Kenyon College had their spring meeting this weekend, at which they decide who gets promotions and tenure. I will receive a letter in the mail tomorrow, telling me either that I have received or been denied "Appointment Without Limit" (the official term for tenure). Until two or three days ago, I did not lose any sleep over the decision, but then it occurred to me that the mysterious meeting [the chair] had with the president and provost may have been for them to give him advance warning that they were not going to recommend me for tenure. That thought caused me to lose, well, perhaps fifteen or twenty minutes of sleep. I'm a very sound sleeper.

The Ax Falls. *As you read the following entries, you will read some very negative comments about Kenyon College and its administration, not all of them supported by factual evidence. To be fair to Kenyon, please bear in mind that many of them were made by friends and colleagues who wanted to boost my spirits, and thus could not be wholly objective.*

4/26/94: The difference between being guillotined and being denied tenure is
that after being denied tenure, you're still alive. However, the sense of incredulity
is the same. You mean this is really happening? You mean there's nothing I can
do?

Although I had some anxiety about the upcoming decision this weekend, I
fully expected to get the letter on Monday morning saying that I had received
Appointment without Limit. Kay was even more certain that I could not possibly
be denied. We planned to meet at the Post Office after my 9:10-10:00 class so that
we could end the suspense together. But right after my class, a very ominous phone
call came from the provost's secretary, saying that the provost wanted to meet me
at 11:00. At that point I felt certain that something was wrong, and the sense of
certainty grew when Kay and I opened the mailbox and found no official letter.
"That's cruel!" she said. But there's no way to take away someone's job without
being cruel.

At 11:00 I arrived at the provost's office, the secretary went across the hall
to summon the president, and he joined us. The meeting was very brief; it was
finished by 11:10. The provost informed me that they had not found it possible to
recommend me for Appointment without Limit; that he could not go into detail,
but the reason was my teaching. There was not much for me to say. I asked if the
criticisms of my teaching came entirely from students or if there had been comments
from faculty. The provost said that the concern came from across the spectrum. I
asked if the financial circumstances or the appearance of P. F. Kluge's book, with
its criticism of the College's recent record of tenuring everybody, had "changed the
rules" in any way, made them set the standards higher than they had been. The
provost said no. Therefore I am left to infer that they consider me not only the
worst teacher of the ten who came up for tenure this year, but the worst to come
up for tenure in several years . . .

It seemed as if I spent most of the afternoon and evening talking . . . First
I talked with Kay and the department chair in my office; the chair assured me
again that I had been given the department's unanimous support. He gave me
a copy of the departmental letter of recommendation, and also told me that in
his own recommendation he had called me "the department's best mathematician
since Nikodym," a staggering compliment (Nikodym was a world-famous mathematician, and retired—ironically, under pressure from the administration—in 1964)
. . . Another professor in the department said she had gotten an inkling of what was
to happen last week, when she was called into a meeting with the president and
provost. Since she was the only member of the department that I had not asked for
a letter of recommendation (the rules required me to ask for four, and there are five
other people in the department), they wanted to find out her opinion. Actually,
that shows they may not have made up their minds even as late as last Wednesday. But she said that each time she told them something positive about me, the
response was, "Yes, we already know that." And the questions they asked her were
things she simply could not answer.

Later I met with a history professor, who wrote one of my letters of recommendation. We sat under a tree in the graveyard (she said, "I hope you don't mind the
symbolism") and talked for over an hour. She was a little skeptical at first of my
theory that the decision might have been dictated by the financial pressures and
extra mural pressures for greater "accountability." But the more she thought about
it, the more it made sense to her—with ten people, an unusually large number,

being evaluated, it may have seemed irresponsible to the president and provost to give blanket tenure to all ten. Of course, this theory is completely unprovable, because they would never admit to it; and, in a way, it is beside the point now. The decision is made.

4/29/94: I've been feeling much better, in fact positively chipper, over the last three days. So many people have told me that they felt the tenure decision was wrong that I have ceased to see it as a personal failure. Kay's boss said it was the "stupidest thing I've heard in ten years." A biology professor brought us flowers and homemade goodies and said, "Of the ten people who were up for tenure I would have put you at the top, not the bottom." An English professor ... was outraged because she thought it was due to the new system whereby students can send in their evaluations by e-mail, which makes it too easy for them to say things that they would not say in a normal letter. A co-worker of Kay said, "It stinks and it's rotten." [Five other colleagues] and probably others I've forgotten have all expressed various forms of dismay or bewilderment. It's especially impressive how many of these people have gone out of their way to talk to us and express their support ... We've come to see that a lot of people do appreciate us. It's not Kenyon that has rejected us, but two people at Kenyon ...

There will be some more interesting developments in the next few days. Today the math department had a meeting (without me) to discuss the decision, and on Monday they will have a meeting with the provost ... On Tuesday the science division will have a meeting, at which one agenda item is a discussion of the promotion and tenure procedure. Two biology professors say they don't think there is a single person in the division who is not upset by the decision, because of its implications for all departments: research doesn't really matter, and the opinions of a few disgruntled students (of which there are always plenty in any intro science course) can outweigh the opinions of the entire department ...

At this point it may be necessary to explain a few peculiarities of Kenyon's tenure review system. At the time of this narrative, Kenyon was practically unique among American colleges in not having a Promotion and Tenure Committee. The decision on whether to recommend a candidate for tenure at the trustees' meeting was made entirely by the provost and president, based on a dossier consisting of the following: four letters from faculty in the department, a departmental letter, three letters from faculty outside the department, two letters from faculty at other institutions, and a minimum of 16 (remember this number!) letters from students, out of a list of 36 students compiled half by the tenure candidate and half by the provost. Unlike many other institutions, Kenyon does not use standardized student evaluation forms. Finally, again contrary to standard practice at other institutions, the candidate's department has no access to the dossier.

5/2/94: The department met with the provost today and got a few answers, though not very satisfying ones. He did give out some information on the student letters: out of 36 requested, only 16 were received (but this is fairly normal, and enough to constitute a dossier); of these, he said that four could be characterized as "generally positive" and 12 were "generally negative." Those are daunting numbers. To put it another way, my approval rating was only slightly higher than Richard Nixon's when he resigned the presidency. It's difficult to comprehend how this could be. For one thing, it's amazing to think that out of the 18 names I gave to the provost as students who I thought would probably give me a favorable review, at most four actually did. Either I am vastly mistaken as to the opinion these students held of me, or the administration is reading the letters in a most unusual way.

This evening I thought of three things that I could have done to improve my chances of getting tenure, if I had thought I was in serious trouble. I plan to mention these at tomorrow's science division meeting, for the benefit of people who will come up for tenure in the future. First, collect student evaluations, whether this is departmental policy or not. One reason is to find out about student dissatisfaction early enough to do something about it. The second reason is more cynical: so that you can defend yourself if the administration tries to say that you have a 25% approval rating ... Second, if there is concern about your teaching, get a senior faculty person to sit in on your course. Again, there is a positive reason—this person can act as a mentor—and a defensive reason—this person can vouch for what happened in the course even if some students say something ridiculous about it ... Third, and something that would never have occurred to me before: get out the vote. If the administration insists on treating the tenure evaluation as a popularity contest, then any faculty member will improve his or her chances by contacting individually the 18 students on his or her list, impressing on them the importance of their letters, and urging them to write.

5/7/94: At the science division meeting it was decided that the division chair would write a letter for the division to the president and provost, but it would not be so much a letter of protest as a letter saying that the tenure decision had raised certain questions and problems about the process ... I also gave my advice about how to improve the odds in the tenure process. A biology professor made a most interesting response to that. She said that, the year she came up for tenure, she told all of her students about it, stressed the important consequences the student letters could have and told them that anything negative they said could be used as a pretext to take away her job. As a result, she said, "that was the only evaluation where I didn't get any negative letters."

I've set up a lunch meeting with the provost for next Thursday, and made up a list of several more questions to ask him. After that, and after I've seen the written explanation of why I was denied tenure (which he says I should receive before our meeting), I'll decide whether I want to press a grievance. At present I think that I probably will.

According to the faculty handbook, there are two possible grounds for a grievance: I could either claim a procedural error by the administration, or I could claim that my dossier was not interpreted in a reasonable manner. I think that my best case for a procedural error was that the administration did consult with the chair and one other professor when it was apparent that my case was problematic, but they did not consult with them in a way which would have allowed them to respond effectively. Neither of them was told ahead of time what the subject of their meeting was ...

Probably my better case is to argue that the dossier was not interpreted reasonably. Here I can bring up the "formula" by which they are supposed to evaluate it: 55% teaching, 30% research, and 15% collegiate citizenship ...

The percentages alluded to above were approved by the faculty, in a perhaps misguided attempt to quantify the unquantifiable. Though it would be impossible to enforce them in any precise way, certainly a gross violation of this policy could be construed as a procedural error.

On May 11 I received the promised letter from the provost outlining the reasons for the negative tenure decision. The letter summarizes my scholarly engagement (30% of the decision) in seven lines and my collegiate citizenship (15% of the decision) in eight lines, and then contains fifty-six lines of commentary on my teaching. Among other things, the provost wrote:

> Looking at student evaluations first, I read fewer than one-fourth that are essentially unqualified in their support of your teaching. All of these are from very able students. That leaves a large majority of letters that are mixed or negative. What are you faulted for? A lack of organization in your presentations, a poor sense of your audience and of their difficulties in comprehending what you are teaching, a tendency to criticize unfairly and thereby to intimidate, a tone of unfriendliness towards many students that makes them reluctant to seek your assistance ... In general, students from upper-level courses can find elements of strength in your teaching that in some fashion compensate for the difficulties they write about; students from the introductory calculus sequence tend to be simply unenthusiastic about your teaching.
>
> There can be no doubt that you are bright, no doubt that you are a very fine mathematician. But you are poorest precisely where the department needs strength—in its introductory calculus sequence ...

5/12/94: I'm beginning to find out that contesting my tenure decision, even if it's the right thing to do, is going to be a little bit time-consuming. Last night I spent the entire evening typing up, re-thinking and re-typing the questions I was going to ask the provost in our lunch meeting today. The meeting itself lasted close to two hours. Later in the afternoon I met with an art history professor who went through the grievance procedure five years ago when he was denied promotion to full professor. That meeting lasted another hour and a quarter ...

The meeting with the art history professor was even more worthwhile than I expected. He expressed the opinion that the provost is really only a front man for the president. He said that I would be surprised how cursorily the dossier is actually read; he believes that most of the alleged "concerns" in the provost's letter are sought out after the decision has been made. He said that the most important part of my defense is to have the department behind me, and their number one question should be why the president and provost overturned the departmental recommendation, which he characterized as a "terrible precedent" and "serious business." The number two question should be whether that has ever happened before. He said that he thought it was likely that I would win the appeal before the grievance committee ... He described to a tee the tactics that the provost has used so far; he said that he would try to "scare the department with the prospect of terrible letters" (i.e. portray the student letters of recommendation, to which we have no access, as extremely negative, so the department will not feel as if it has a chance to win), but that the department should not be convinced. He thought that I had some very good arguments that procedural violations had occurred, starting with the argument that worked for him, namely that the provost's "summary" of the dossier was not a summary but in fact a highly slanted rationalization for his own action ...

After my meeting with the art history professor, I'm not sure just how important the accumulation of arguments and counter-arguments will be, if the decision was actually made independently, or somewhat independently, of the facts in the dossier. My conversation with him brought me back to one of my first reactions to the decision: that the administration had been looking to deny tenure to one of the ten candidates, and I was the easiest target. As he said, "It's the same way that a mugger thinks." But, of course, such assertions can never be proved, which is why one has to expend so much time and energy trying to catch them in a procedural error.

5/16/94: Driving to Columbus on Saturday gave me some time to ponder the case a bit more. I'm almost sure that I will file a grievance now. That day was the first day that I felt 100% certain that I was in the right and that I ought to be able to win my case before the grievance committee.

5/22/94: The battle over my tenure decision continues to simmer. I had meetings with the chair of the faculty, who thought I had a strong case and should go ahead with a grievance ... The chair of the department and the division had an unproductive meeting with the president and provost, during which, as the department chair reported, the president did "90% of the talking."

Kay and I had a friend over for dinner on Wednesday night, a great morale booster because she has a very low opinion of the president and provost ... She considers the president to be like King Lear, receiving counsel from all the wrong places and ignoring most of it ...

In retrospect, the administration's arrogance at this stage of the procedure was breathtaking. According to the department chair, during his harangue the president said that the decision could not be reconsidered unless they provided "proof of error." Of course, since the department had no access to the dossier, it is difficult to imagine what such "proof of error" could consist of. As we shall see, the dossier did contain proof of error that the president must have known about.

5/25/94: The wheels have been set in motion. On Monday I delivered my grievance letter to the provost's office. Tomorrow I will have a meeting with the chair of the Grievance Committee. I don't know what to expect from this meeting ...

Following the advice of the chair of the college's Grievance Committee to be as specific as possible, I itemized six procedural errors and seven errors of interpretation. I argued that the administration had unreasonably overruled the department's professional opinion on flimsy evidence—the minimum allowable number of student letters. Some other possible errors I cited were the meetings with two members of the department who were not given any advance knowledge of the agenda; failure to interpret my mentorship of Kenyon Summer Science Scholars (a summer research program for undergraduates) as teaching; failure to take into account specific ways that I had improved since the second re-appointment; and a summary letter from the provost that was not a fair representation of the dossier.

Grievance. *The section of Kenyon's faculty handbook dealing with the grievance procedure is one section I had never thought to look at before April 1994. Kenyon's grievance procedure has three stages. First, there is a relatively amorphous stage called "Informal Consultation," in which "the president or provost will seek to resolve the dispute informally by consultation with the faculty member, the faculty member's department chair and others whose knowledge or experience may* *be of help … " In essence, this phase had already concluded by the time I wrote my grievance letter. The second phase is "Mediation" by a mediator appointed by the chair of the Grievance Committee. The final phase, which the faculty member may invoke "in the case of failure of other efforts to resolve the dispute," is the appointment of a Hearing Panel consisting of three people from the college's standing Grievance Committee. The panel decides "whether the evidence warrants a grievance hearing," and if so, the case goes to a formal hearing. This last step is not taken lightly by anyone; for the hearing panel, it means hours of difficult and sometimes emotionally charged work. After the hearing ends, the panel makes a written statement of its findings to all parties, and the president of the college is required to accept or reject the recommendation of the panel within one week. The scope of the panel's authority is strictly limited: it is only allowed to recommend a re-evaluation of the tenure candidate, not to conduct its own evaluation, and its recommendation is not binding on the president.*

5/28/94: The latest twist in the saga of my tenure review is not an encouraging one. Having filed my grievance on Monday, my next step was to meet with the chair of the Grievance Committee, for the "informal consultation" phase of the process. We met on Thursday, and then he met with the provost on Friday, and was allowed to look at the dossier. Today I got an e-mail from him (actually sent last night) in which he stated that he did not think that the provost had interpreted the dossier "wrongfully." He advised me not to pursue the grievance further, although he added that it was, of course, my choice. I was dismayed not only by his conclusion, but also by the way he drew it. Though he advised me to put as much into writing as possible in my grievance, and Kay and I slaved over it last weekend, it really seems to have made hardly a bit of difference. He barely even referred to it; only in the postscript did he address anything I wrote in the letter … the arguments in my letter have been not so much answered as simply brushed off.

So, sooner than I expected, another crossroads is reached. I certainly don't want to go out just looking like a sore loser, someone who can't face up to reality … It would be foolish to be optimistic at this point about the result of the grievance procedure. But on the other hand, I do think it's reasonable to expect some answers.

Next week I plan to talk with the chair of the faculty again, and see what he makes of it, and whether he still advises me to continue. I may also talk with the art history professor and the history professor again. I think the only certain thing is that if no one advises me to continue, I will not. One reason is that I am entitled to have a faculty advocate in the grievance procedure, and I want someone who's at least somewhat enthusiastic and thinks I can win. I'm pretty sure that the faculty chair is the man I would want for that role … I think it's important to have a more senior, more well-connected member of the faculty who is willing to defend me.

This was probably the lowest moment for me since the day I was first informed of the tenure decision. An independent, presumably unbiased reader had looked at my dossier and found nothing to contradict what the provost had written. However, my conversation with the chair of the faculty did much to lift my spirits. I did, in fact, choose him as my advocate. A roly-poly religion professor with thick glasses, resembling Santa Claus without the beard, he was an ideal choice for the role: good-humored enough not to offend anyone but savvy enough to know what the important issues were. While I often became bogged down in a morass of arguments and counter-arguments, he constantly advised me not to be too "legalistic." I will refer to him below as "Len" (not his real name).

For anyone who finds him or herself in a similar position, I cannot overempha-size the importance of finding a senior faculty member to act as your advocate. As my case shows, it need not be someone from your own department.

5/31/94: Yesterday morning, even though it was Memorial Day, I met with Len to discuss the latest development. He also found the Grievance Committee chair's response to be a little bit puzzling, because it wasn't clear whether he was interpreting this phase as the "informal consultation" or the "mediation." In the "informal consultation" phase the dossier is still supposed to be closed, and the chairman of the Grievance Committee is not supposed to be involved. So it seems likely that he was acting as a self-appointed mediator, which Len called "irregular," though not necessarily illegal ... He also said that the Grievance Committee chair's reaction seemed a little bit impatient to him ... Finally, he said that he felt "no less strongly than before" about the validity of my case, and in addition felt a certain amount of dismay at the hasty and cavalier way the Grievance Committee chair had dealt with my petition.

In other words, Len gave me precisely the support I was looking for to justify continuing my grievance ...

On June 18 I wrote a letter to the Grievance Committee chair restating the complaints that I felt had not been answered from my initial grievance letter, and requesting a hearing. On June 30 the Grievance Committee replied that a hearing panel would be formed in late August. Over the summer some changes took place in the administration: a new provost took office (however, the old provost would be required to defend his own decision in the hearing), and the president announced his resignation, effective at the end of the following school year, after twenty years in office, the longest term of any active college president in the country.

According to the rules of the grievance procedure, I was not allowed to see my own dossier until ten days before the grievance hearing. On September 12 I finally got to see with my own eyes the evidence that had led the administration to deny me tenure.

9/12/94: Wow! ... Today, with Len, my faculty advisor, I finally got to view the contents of my dossier in the provost's office. I think it is fair to say that we were both astonished. The student letters, which the provost had led us to believe were mostly negative, were in fact mostly positive; and the faculty letters, which had been portrayed as ambiguous, were overwhelmingly clear in their support of my candidacy for tenure. The impression that we both got from the dossier was so dramatically different from the tone of the provost's letter that it is virtually impossible to imagine any more that the evaluation was conducted in good faith. One would in fact have to read the letters with careful attention to all negative comments to construct a summary as negative as the provost's. A number of writers, while generally praising my teaching, would write their letters with a sentence beginning "His greatest strength is ... " and another sentence beginning "His greatest weakness is ... " This is only a sign of a person attempting to give an objective and balanced evaluation. But every time, the provost reported only the negative comments and interpreted the letter as showing a "mixed" opinion or worse.

Some more important discoveries: there were at least two blatant procedural errors. First, instead of the minimum of 16 student letters, the provost received only 15, one of which simply said that the student could not provide any information ... The provost stated in a letter to the president, in fact, that he "could not secure sixteen student letters." Yet he repeatedly told me, "the dossier is complete and adequate to its purpose," even though I specifically asked about the number of letters received.

The second blatant error is that no letter was received from the faculty member outside my department who was supposed to evaluate my teaching. Since the decision was purportedly based on teaching, one is amazed that the provost and president did nothing to rectify this omission. It's even more amazing in light of the fact that that faculty member says he did send a letter . . .

In short, the administration's case seems to me quite a lot worse than I even suspected . . .

9/17/94: On Thursday I talked with the author of the mysterious disappearing letter: his evaluation of my teaching, which he says he sent in early January but the Provost's office apparently never received. The letter itself, as Len observed, is not going to blow the lid off the case . . . The things that make the letter more important are that it was presumably sent yet did not appear in the dossier, that the administration felt comfortable in making a decision based on my teaching even in the absence of a faculty letter from outside the department on my teaching, the fact that I was never informed that the letter was missing, . . . and the fact that the provost misled me after the decision by saying the dossier was complete.

Yesterday, Friday, I had my interview with the grievance panel. It lasted about an hour and a half, and went pretty well. There is no question that they are taking the case seriously, and on some points of substance I think they already agree with me. They had already considered and basically ruled out my suggestion that the administration might not have acted in good faith; however, the chair of the panel did say that he felt that after the decision was made, the provost's letter had been constructed in such a way as to justify the decision rather than to reflect the dossier . . .

There were too many interesting details covered in our meeting to recount them all. I will just mention one more thing. Apparently, when they talked with the president, one of the arguments he had considered most important was as follows: if I turn off students in lower-level courses so that they never took a math course again, then it doesn't really matter how good a teacher I am for the upper-level courses. So one of the panelists asked my chairman to study the validity of the president's hypothesis: do I in fact turn off the introductory students? The chair identified all of the students who have taken me for their first math course, and computed the average number of math courses they have taken after that. For comparison, he did the same thing with another professor in the department. The result was striking: my students have averaged 1.1 more math courses, and that professor's have averaged 0.8. Yet I am the one who is supposedly depressing math enrollments?

9/20/94: Thank goodness the hearing is over. It was just enough to get me heartily sick of this whole tenure controversy again. Having said that, though, I should also say that I think the hearing was quite productive in some ways. Once again, I could write a very long entry describing all the details, but since many of the details will be made moot by the grievance panel's decision, I will try to give a condensed version.

There were two particularly encouraging things about the hearing. First was the testimony from the members of my department. I think the panel had some serious doubts about the strength of the department's support (one panelist seemed to think that their letters were "mixed"), and I believe their testimony should convince the panel that their support was in no way mixed. I think the panel will have to decide whether it is reasonable to believe that there could be a serious problem with my teaching, as the provost and president allege, that no one in my department perceives. This was implicit in a question one panelist put to the president, about how he could account for such an apparent dichotomy between the faculty and student views of my teaching. The president, as he did throughout the hearing, essentially stonewalled the question, saying that they did not perceive a dichotomy. But I doubt that his answer will persuade the committee. (Although, incidentally, I might agree with the president in another way: there wasn't so much of a dichotomy because in fact the student letters weren't all that negative.)

Getting back to the main point, another very helpful part of the department's testimony was that it revealed some specific ways in which the administration misunderstood what they had said. For example, there was a sentence in one professor's letter that mentioned that the class he had observed had "started late" because some students straggled in late (it was an 8:30 class) and I was still collecting homework from them up to ten minutes after the start of class. But the president and provost had interpreted "starting late" to mean that I had actually come to class ten minutes late—which, as the faculty member said, was not true. Moreover, this single comment, the only negative sentence in that letter out of three pages of glowing positives, was the only thing that the provost had cited in his letter to the president recommending that I not be tenured. The professor told the hearing panel that he felt he had to include something negative in the letter or else it would not be taken seriously. Instead, the negative comment was the only thing that was taken seriously.

Another miscommunication was apparent when the second member of the department testified. The chair of the panel asked her to clarify her "now famous comment" (only to the panel, of course) that she agreed with the administration's decision—something that was cited by both the president and provost when they met with the panel. She was shocked, and said that she had never said such a thing. What she had meant was that they had access to the dossier and she didn't, so she could not know what was in it, but if indeed the letters from students were as negative as portrayed by the provost then she could understand the decision. That's a lot different from agreeing with it! The third department member's testimony was also helpful. We discussed the fact that he had essentially written the department's letter of recommendation. The chair of the panel asked him the question I had wanted to ask but didn't quite know how: did the strength of the department's letter have something to do with who wrote it? The poor guy thought and thought and finally said, "Perhaps I understate things." His comment was so ingenuous and so ... well, understated, that I do not think he could have possibly given a better answer ...

Len didn't say a whole lot, but what he did say was very helpful. While the administration kept harping on negative student letters, he reiterated that he had not found the letters to be negative at all, and that he personally would have been happy to come up for tenure with such a dossier. I think the panel has to take it very seriously when a respected senior faculty member like him says that and means it. He also provided one of the few moments of comic relief, when he asked the president whether anyone actually has a dossier with no negative letters at all. The president said yes, and Len asked, "And you believe them?" Everyone laughed, but I think that part of the reason for the laughter was that it was a point well made.

9/22/94: Some of the suspense ended today ... In a very well-written and well-reasoned letter, the hearing panel gave me virtually a complete victory. They argued that the administration had not followed the proper procedures by failing to notify me that my dossier was incomplete; that this may have adversely affected the quality of my dossier by depriving me of a chance to solicit letters from students; and that the student and faculty letters in the dossier had been misinterpreted. Accordingly, they recommended that I be re-evaluated for tenure.

This victory means a lot to me, both as a moral victory and as a decision that will wipe the "black mark" off my record if I apply to other institutions for a job. Now, instead of giving my personal opinion that the tenure decision was misguided, I have an official determination from a faculty committee that was able to examine all the evidence.

Some of the passages from the Grievance Panel's report were quite tart, and I read them with a great sense of vindication. A few of them are given below:

> The Faculty Handbook states that during the evaluation for appointment without limit: 'By January 31, the Provost will inform the faculty member which materials and letters from the evaluators chosen by the member have not been received.' By the Provost's own admission, one faculty letter and several student letters remained outstanding at this time. Yet Mackenzie was never informed; and the dossier remained incomplete when the decision was made.
>
> It must be emphasized that all persons evaluated deserve at minimum a dossier compiled according to our basic regulations. That the rule regarding notification is routinely ignored, as the Provost testified, does not in any way excuse this lapse ... The failure to notify was particularly serious given that the missing items related specifically to teaching, the area where deficiencies proved decisive in the review ...
>
> The central reason for denying tenure to Mackenzie was his performance in teaching introductory Calculus, and the main evidence for his inadequacy in that area was the student letters. But the Provost's interpretation of that evidence ... seems to us in several respects an unreasonable representation of the student letters. The provost claims that 'Students from the introductory calculus sequence tend to be simply unenthusiastic about your teaching.' We found however that some of those students were in fact extremely enthusiastic. The provost writes that there is 'a large majority of letters that are mixed or negative.' Although there clearly are letters that are mixed or negative, they do not in our view constitute a majority, let alone a large majority ...
>
> Whatever strengths [students from upper-level courses] saw were wrongly interpreted by the Provost as mainly or merely compensation for weaknesses. On the contrary, our sense was that in the main the advanced students saw Mackenzie's teaching as exceptionally positive ...

Given the administrators' acknowledgment of the standards and candor of the Math faculty, it is particularly disturbing that the Chair ... was invited for a critical meeting without knowledge that the subject concerned an impending negative decision on Mackenzie. This ignorance was intended, the Provost states, to prevent [the Chair] from somehow making inappropriate preparations for the discussion. As a result [the Chair] felt he inadequately defended Mackenzie's record, and the administrators incorrectly inferred that he did not significantly dispute their conclusions ...

9/30/94: [In] my mailbox I found a letter from the president that was as welcome as the letter that I waited for in vain on April 25. In five terse lines, the president acknowledged the grievance panel's recommendation that I be re-evaluated for tenure and said that he accepted the recommendation.

To return to the guillotine metaphor I used last April, I guess I feel now like someone whose head has been sewn back on: giddy with relief, but still in somewhat precarious health.

Double Jeopardy. *After accepting the results of the grievance hearing, the president wrote another letter outlining the procedure that would be followed for my re-evaluation. Since the administration did not dispute my qualifications in scholarly engagement and collegiate citizenship, the new review would focus exclusively on my teaching. And the scrutiny would be more intense this time. Every student whom I had taught in the last two years would be asked to write a letter, and the faculty in my department arranged to attend several of my classes, where in past reviews they had only attended one or two. I entered the new review cautiously optimistic: cautious because I knew the review would be conducted by the same president, but optimistic because it would be conducted by a new provost, and because I felt that my department's support would be much more clearly expressed this time.*

One complication that arose in the fall was the department's use of a new "reform" calculus book, Calculus in Context. As we expected, the new and radically different approach to calculus drew a lot of criticism from students (in fact, the department abandoned this book two years later); however, my colleagues pledged to keep student criticism of the book as separate from their evaluation of my teaching as possible.

Here are a few of my teaching experiences from that fall.

9/23/94: There were two interesting points in today's class. First, when I discussed reaction rates as an example of exponential growth or decay (reaction rate is proportional to the concentration of the reactant), one student said, "That's not the way we do it in chemistry!" But then he thought about it a bit and said, "Wait a minute ... there's something about taking the logarithm ... maybe it is the same thing!" He said that in the chemistry course they just learn a rote technique for finding the reaction rate, without learning why it works. Now he might understand why!

The second point came up when we were discussing inverse functions. As usual, this provoked a certain amount of confusion among the students. I think my way of explaining it is partly at fault. The book has a very nice way, which I will try on Monday. Anyway, another student came up to me after class and started explaining how he learned about inverse functions in high school. To paraphrase: "An inverse function is ... you switch x and y, and then you solve for y."

To me, this was another perfect example of how students are taught rote procedures for getting the right answer, without really understanding the concepts involved.

11/11/94: One of my colleagues has started sitting in on my class, and the experience has already been beneficial to both of us. First, she really liked the way I used DERIVE to explain why an unbounded region can have such a narrow "neck" that it has finite area. When you plot a function like $y = (1 - x)^{(-1/2)}$, DERIVE cannot even show the asymptote ... the neck is so narrow that the computer can't even "see" it. My colleague said she will always introduce improper integrals that way from now on ...

If teaching is a battle for souls, I won one and lost one this week. (Perhaps.) One of the students who has been most critical of the Calculus in Context approach wrote in his journal that he had been thinking about some other subjects over the weekend, and suddenly this approach started to make sense to him after all. He was very vague about it, and wrote, "I will have to think more about this," but that was a very encouraging sign indeed!

The setback occurred this morning. One of my students asked if she could have 5 minutes after class to do a little computer work for one of the problems on the take-home exam, because she hadn't had time to come to the computer lab last night. I said no, and explained, "You're supposed to make time to come to the lab." She got upset, said, "You shouldn't say that, because I worked on this test for nine hours yesterday," and stormed out of the classroom in tears.

There are so many aspects of this incident that I can second-guess myself on. Was it unreasonable to deny her the five minutes? No. A deadline is a deadline. Another student had the same problem on the last test, and lost several points as a result. I have to be consistent. Was my comment insensitive? I don't know. At that point I couldn't have known how much time she had put into the exam already. Some students need a little lecture like that to get the message. Was the exam too long? Apparently most of the students took a very long time to do the first problem. I was very surprised, because the book shows, step by step, how to solve this kind of problem (a logistic equation) and even gives a formula for the solution. BUT ... there was only one homework problem on the logistic equation, and it had a typo that ruined the problem, so I didn't count that homework problem. And so, the students, minimizing effort as students always do, may have thought, "Well, the homework problem didn't count, so we won't be responsible for this on the test."

11/16/94: I felt lower than low after this morning's class ... I guess I should have stayed away from the [problem] that caused all the emotion on Friday. One student had gotten an unrealistic answer and seemed puzzled about it, so I had written next to it, "Garbage in, garbage out"—meaning that because the equation he had plugged some numerical values into was wrong, the output was also wrong. But he interpreted it, I think, as a comment on his whole solution, and started telling me how long he had worked on it, etc. I got pretty flustered, partly because I knew I had set myself up by writing a comment that could be so easily misinterpreted. Ordinarily I would have patched it up and moved on, but after all my tenure struggles I have gotten so paranoid. "Is this where I lose the student forever? Is this what he's going to write about in his letter to the provost? What are [the two math professors attending my class] going to think?" For about 15 minutes I felt as if my brain was disconnected from my mouth, as I babbled on about that problem . . .

Every semester has to have a worst class, and I hope this morning's class was it.

Perhaps I was right to be so paranoid: one of the two professors in attendance told me, months later, that this class had made a big impression on him. As for the student, my apprehensions were wrong: I didn't "lose" him, and perhaps he even forgot all about the incident. The last time I talked with him, over a year and a half later, he commented on how much he had learned from my class.

12/1/94: . . . Another highlight yesterday was my morning calculus class, which was visited by the chair of the math department and the new provost. I started the chapter on dynamical systems . . . The timing was fortuitous, because dynamical systems is an area of mathematics the provost knows a lot about, and I think he was probably pleased to see it covered in a calculus course. The chair was also very excited about my class, particularly about the way I pointed out that the computer's drawing of trajectories "slows down" as they approach equilibrium points. He thought it was neat that you could actually get information not just from the curves themselves, but also the way that the computer draws them. Funny, it seemed sort of obvious to me, but I guess it wasn't. Moreover, it wasn't obvious to the students either, since we had never talked about parametrized curves before. One of the students asked me to explain what the chair meant [and why it was so exciting]. Once again, it was a case where having another faculty member attending my class was a help to me and my students and the other faculty member.

12/29/94: [A former student whom I visited with during Christmas break] paid me a compliment that I never expected to hear. She said that, as she was preparing for her student teaching, she looked over her old tests from the calculus course she took from me, and appreciated for the first time the creativity and wit that went into them.

1/29/95: Friday was the day that the student and faculty letters of evaluation for my tenure review were due at the provost's office. The provost's secretary reported to me on Friday afternoon that they had received 36 student letters and all the faculty letters. Quite a change from last year! Lack of information should not be a problem this time.

As I awaited the outcome of the review, an interesting subplot played itself out: the faculty debated and finally adopted a proposal to create a tenure and promotion committee—too late, ironically, to have any effect on my case.

3/31/95: Three people have told me this week that they were glad that I spoke up in the faculty meeting on Monday . . . I was the first person to speak in the debate on the tenure and promotion committee. I said that I had a unique perspective on the current tenure system, having become the answer to the trivia question, "Who was the last person to be denied tenure at Kenyon?" Then I talked about my view that the departmental input was not great enough, and asked how the proposed committee would affect that; also, I said that a paramount consideration should not be whether more or fewer people get tenure, but whether more or fewer mistakes will be made. I don't think that my little speech was very eloquent, but I guess some people may have thought it was brave for me to identify myself as a person who didn't get tenure.

Finally, four days before the trustees' meeting, I got a hint of the way the wind was blowing.

4/17/95: Once more the same nightmare? Only a nightmare the second time no longer makes the pulse race quite as much ... I got a call from the provost's secretary, who had been told to set up a meeting for me with the president and provost on Thursday. The agenda: my tenure decision. Naturally, two possibilities crossed my mind. One was that they may have decided, out of sympathy, to end my suspense and let me know before the meeting that they were recommending me for tenure. However, that doesn't seem likely, as sympathy is a foreign concept to bureaucracies. The alternative explanation is that I am being denied again. Further support for that interpretation came when a German professor met me in the copier room a few minutes later and asked if I had gotten a call to meet with the "diumvirate." I said I had and, with my hopes momentarily rising, asked if everyone who was up for tenure was getting such calls. She said they definitely weren't. So it almost certainly seems to be bad news for both of us. She was distraught, and looked just the way I remember feeling last year: like a tree uprooted. I felt a lot calmer, since I've been through it before and was somewhat prepared.

4/21/95: After all the surprising turns that my tenure saga has taken, one more shocker awaited me on Thursday morning. As I expected, the president told me that I would not be offered tenure. But there was one huge difference from last year: this time the mathematics department recommended that I not be offered tenure. Once I heard that, the wind went right out of my sails. All that I battled for in the grievance procedure last year was the right to be judged by my own peers. Now that has happened ...

Since Thursday morning, I have talked with each of the members of the department to find out what caused them to change their minds. I think that [one of them] expressed it best. She said that she went into the re-evaluation determined to find the answers to two questions. First, was there a problem with my teaching, or was it a figment of the administration's imagination? And second, if there was a problem, how could it have escaped the department's observation for so long? She said that after sitting in on eight of my classes, she felt that she had the answers. She saw patterns in my teaching that, in individual classes, had not seemed like serious problems, but when they were repeated she could understand why the average to weaker students were dissatisfied. She commented, for example, that I would give a beautifully prepared lecture with nice examples, get to the end, and she would think, "Great, now all he has to do is tie this up"—and instead I would go on to the next topic. She also commented that when students ask questions, she always tries to figure out what it really is they don't understand—which is not always the same as the question asked, because students often don't realize quite what they are confused about. But she said that too often I would take the question too literally, and answer only what the student asked. Another criticism she had was that, because of my mild-mannered demeanor, it was hard to tell the central points of the lecture apart from the minor points. They were all presented on an even keel. Another colleague saw some other problems, such as my not getting all the students equally involved. Also, he pointed out that I would often ask a question, get a right answer, and then go on with the lecture without making sure that everyone understood the answer.

Maybe none of these problems individually was decisive, but taken all together, they made the department too uneasy to recommend me for tenure. My reaction to them was that all the criticisms had some validity, but it was a shame that no one had brought them to my attention four years ago, or even two years ago. It was a fault that we all shared. I did get a warning, in my second re-appointment review, that I should find a mentor to work with me on my teaching. The chair and I talked about having him attend my classes, but we never quite found the time, and I don't think that either of us really believed it was serious enough to warrant the effort. We have all learned that attending each other's classes and talking about them should be a routine part of our business. It should start the first year that new faculty come in, and it should continue even with the senior faculty, because they, too, have to deal with the same kinds of classroom challenges the junior faculty do.

In the above entry I portrayed the math department's change of heart in probably the most favorable light. Other people, including my wife, were not so charitable in their opinion of the department. My wife found support from a somewhat surprising source.

5/6/95: Kay went to the college's ombudsman to talk about my tenure decision and her anger over it. Surprisingly, even though the ombudsman is in the administration, she agreed that I had been badly treated. She had also talked with the German professor who was denied tenure, and agreed that the secrecy of the meetings between the administration and the departments was a serious problem. As the German professor commented on Saturday night [when she visited our house for dinner], the secrecy works completely against the tenure candidate, by depriving that person of the ability to defend him or herself before the decision is announced. I have also commented before that the fact that the department cannot view the candidate's complete dossier was a critical factor in my case. If the department had known how exaggerated were the administration's claims about the number of negative student letters in my [previous year's] dossier, they might have reached a different conclusion.

Incidentally, my colleagues in the mathematics department were also very distressed about the secrecy issue. In mid-January, when they made the decision not to recommend me for tenure, they had intended to inform me immediately, but the provost directed them not to. This resulted in three very awkward months for them.

Was the department's change of heart justified? I have talked with colleagues who called it "criminal" and "immoral" to support me one year and recommend against me the next, without giving me a clue until the day I met with the president and provost. A year after the decision, Len told me that the department's flip-flop was, to him, the most surprising aspect of the whole case. One could, of course, put a very simple interpretation on it: when my colleagues actually took the trouble to attend my classes, they found them unsatisfactory.

On the other hand, from my previous experiences I have learned that things are not always so simple. The jury's decision depends on the charge given to the jury (in this case, my departmental colleagues). In this case that charge was (to paraphrase): we have already found Mackenzie's research and collegiate citizenship to be satisfactory, but if his teaching is not up to snuff then he should not be recommended for tenure. Moreover, crucial information was withheld from the jury: the actual contents of the student evaluations. Only one person in the department, the chair, ever heard Len's crucial comment that the letters were positive enough already for me to get tenure, and that he himself would be happy to come up for tenure with such a dossier. The rest of the department was left with the belief that the students were very critical of my teaching. (Note that they also did not get to see the grievance panel's finding that the administration had misrepresented the student letters.) Even the chair never got to see the students' letters, and may have dismissed Len's observation as a rhetorical flourish. Finally, although the mistakes made in the first evaluation were the administration's, it was I who was subjected to increased scrutiny of my teaching. One time this scrutiny clearly affected my teaching was the dreadful class I described on November 16. To summarize, I believe that the unavailability of key evidence, the changing of the rules of evaluation, and the shifting of the burden of proof were more than enough to cause fair-minded people to make the wrong decision.

I will end with the story of another individual who was forced to turn his back on a career he had given his heart to.

5/10/94: When Michael Jordan decided to return to basketball this winter, after spending the last year as a minor-league baseball player, his basketball coach, Phil Jackson, said, "Michael Jordan didn't fail baseball—baseball failed him." I can say the same thing about academia. The only way the analogy breaks down is that I can't go back to being the world's best basketball player, as Michael Jordan can!

Epilogue. *The German professor mentioned in the last two entries won a more satisfying victory than I did. After the "informal consultation" phase of the grievance procedure, the administration offered her a re-evaluation similar to the one I underwent. In the re-evaluation, which was conducted this time by the brand-new Promotion and Tenure Committee as well as a brand-new president and provost, she received tenure. She benefited not only from my experience, but also from having a well-organized team of faculty advocates from other departments. Again, this shows the importance of having someone else to argue your case. At a liberal arts college, it may be harder for a mathematician to mobilize this sort of support, since there are fewer other disciplines that "speak the same language."*

Len, who did such a marvelous job as my advocate and taught me that a few well-chosen words can be more effective than pages of arguments, received one of the two Trustees' Distinguished Teaching Awards in 1996. Ironically, a mathematician won the other one—a vindication for him, as he had been distressed by receiving criticism on his second re-appointment review (in 1993) quite similar to the criticism I had received on mine.

Conclusion. For the person facing a tenure decision or the person, like me, in the uncomfortable position of challenging a tenure decision, here are some final words of advice.

1. Long before the tenure decision, you should make a concerted effort to receive mentoring from other faculty in your department, and to find out what their expectations are. If there is no mentoring system in place, appeal to individuals to help. Also, suggest that department ought to implement a regular system of mentoring and evaluation.

2. Remember that the actual reasons for the tenure decision may be different from the stated reasons; and remember that the perception of reality by the decision-makers is more important than the reality. If there are honest and ethical ways for you to tilt that perception in your favor, by all means do so.

3. Do not assume that administrators know their jobs well, even the purely administrative parts. If they are capable of bungling a decision, they are also capable of bungling the procedures that they are ostensibly supposed to follow.

4. If you fight a tenure decision, expect it to cost you a great deal of time and emotional energy. And then expect it to cost even more than you expected.

5. Do not venture into the fray alone. You need an older, wiser, and better-connected advocate. In a small college, this may mean going outside your department.

6. Watch out for changes in "the rules of the game," whether overt or hidden. If the new rules are set by the administration, they are unlikely to favor you.

7. Watch out for excessive secrecy. Some secrecy is, of course, required to protect the confidentiality of evaluations. But too much secrecy serves as a cover for incompetence or worse. It never serves you, the faculty member being evaluated. Also, question the need for any secrecy that is imposed on you personally. For example, I believe that it was a mistake for me not to show my colleagues the text of the grievance panel's findings, even though it was marked "Confidential." The result was that the administration's interpretation of the dossier was the only official version they ever heard.

Do you know a colleague who was just denied tenure? It's one of the most shattering experiences one can have in academia, and your colleague would greatly appreciate any words of support you can offer, even if you don't know anything about the specifics of the case. Don't treat that person as if he or she had a contagious disease. Also, unless you know something about the case, go lightly on the "Those bastards, they don't know what they're doing" type of comment. Try to accentuate your colleague's positives rather than the administration's negatives.

Are you conducting a job search, and have applications from people who were denied tenure? In today's competitive job market, I know that there is a strong temptation to pass over any candidate who has any negatives on his or her record, such as an adverse tenure decision. Try looking at that candidate differently: this may be your chance to profit from another institution's huge mistake. You may be getting a very experienced professor who just didn't fit in that other place, or who was denied for reasons having little to do with his or her qualifications.

References

1. E. Aboufadel, *The New Job Diary*, MAA FOCUS **13** (1993), no. 5, 6; **14** (1994), no. 1, 2.
2. E. Aboufadel, *The Search Committee Diary*, MAA FOCUS **14** (1994), no. 5, 6; **15** (1995), no. 1.
3. E. Aboufadel, *The Tenure Review Diary*, http://www.maa.org/features/ed/diaries.html, MAA Online.

CHAPTER 8

The Active Mathematical Community

There is more to being a mathematician than solving hard problems and publishing the solutions. The YMN is just one example of "extra-curricular" activity that can engage the attention and the time of mathematicians. In this chapter, we metaphorically open the door of your office and lead you outside. We introduce you to two of our professional societies (the AMS and the MAA). We offer advice on how to afford mathematical meetings, and once you are there, how to give talks or host special sessions.

Both of the editors of this book believe in supporting the mathematical community, not only by giving our time to it but also by joining professional societies (such as the AMS and MAA, although there are many others). We welcome you—and encourage you—to do the same.

Getting Involved with the American Mathematical Society
by Mark W. Winstead

Since a few of my accomplices and I started YMN in the summer of 1993, I have heard numerous comments about how the AMS (or MAA or SIAM) doesn't seem to care, or only seems to listen to senior mathematicians, or otherwise is unresponsive to the wishes or needs of junior mathematicians. Those who comment are thankful for YMN. Interestingly, I have heard just about as many comments from "insiders" grateful to see junior mathematicians getting involved and making their viewpoints known. This latter group usually consists of more seasoned mathematicians who realize that threats that face junior mathematicians put the whole of mathematics at risk, thus it is in the best interest of a subject they love to help junior mathematicians. This group also wishes to see junior mathematicians more involved.

Both of these sets of comments help me realize that there is a bit of a communications gap. Junior mathematicians don't realize how easy it can be to get involved (boy, I found out!) and seasoned mathematical "insiders" don't realize that junior mathematicians don't know how to get involved, other than to "whine" in the pages of *Concerns*.

I would like to outline for you a handful of ways that you can get involved. In particular, I am focusing this article on the AMS.

Suggest Candidates. One way to influence AMS policy and actions is to nominate people for office in the annual AMS elections. Send in your suggestions of good candidates for the offices of Vice- President, Members-at-Large of the Council (up to five) and Member of the Board of Trustees to *AMS Nominating Committee, P.O. Box 6248, Providence, RI 02940*. There is a form for this in the September issue of the Notices. Please note that suggestions are requested to be sent by early November.

Familiarize Yourself With The Nominating Committee. Also in the September *Notices* is a list of AMS officers and committee members. In particular, there is a list of members of the Nominating Committee. If you happen to know one of these people, take the time to tell them about the kinds of candidates you would like to see nominated. If you don't know anyone on the Nominating Committee, you still influence them by getting their e-mail addresses from the Combined Memberships Listings (CML), which is available in paper and also at `www.ams.org/cml`.

Familiarize Yourself With The Members Of The Council And AMS Committees. The Council is required by the AMS bylaws to approve the Nominating Committee's choices, so you may want to have discussions with anyone you know personally from that list, as well as the winners of this fall's elections. Similarly, if you don't know anyone from the list from the September issue of the Notices, you can get e-mail addresses from the CML.

You may also wish to discuss issues with Council members or with members of appropriate committees. Again see the September issue of the *Notices* the list of Council members names and lists of committees and their memberships. Learn who in your department is on the Council or any committees, occasionally ask them about issues facing their committee(s), and offer an opinion.

If you know someone on a committee, you may wish to occasionally point out relevant articles in periodicals. In the case of electronic periodicals, such as *Concerns of Young Mathematicians*, you can clip the articles and send your clippings by e-mail to the committee member(s) you know. Even if they subscribe, they may not have the time to do more than skim the periodical and could have missed an important article.

Use The Petition Option For Placement On The Ballot. Another way to get people on the ballot for next fall is by petition. For offices other than positions on the Nominating Committee, it takes is 50 verifiable signatures to get on the ballot. For the Nominating Committee, it takes 100 verifiable signatures. In the September issue of the Notices, there is be a form for use in obtaining signatures. If you wish to get someone on the ballot by petition, I would suggest you or an accomplice go to the winter meetings and place your petition in the place(s) made available for such things.

VOTE. Enough said.

Participation in the AMS
by Jean Taylor

There are many ways to participate in the AMS in addition to the three that Mark Winstead listed (nominate people for Council or the Nominating Committee, vote, and communicate with people on the Council or Nominating Committee). In particular, if you get hold of the Mathematical Sciences Professional Directory (your department office may have one, or you can contact the AMS) you'll see that there are 12 pages of committees listed, and a two-page index of committees after that! Now a lot of these are specialized, like committees of the Board of Trustees, or editorial committees of journals, but there are quite a few that should be open to young mathematicians. The Committee on Committees makes recommendations to the President of the AMS, and then she appoints people to these various committees. So people who are interested should contact the Committee on Committees presumably through the Secretary), saying they are available.

Also, there is the opportunity to organize Special Sessions for regional and annual and summer meetings. Reasonable proposals are rarely turned down. It is, of course, work to round up people to talk in your Special Session and to schedule who talks when, *etc.*, but most of the organizational details are handled by the AMS meeting staff, so it's not too bad. And the AMS is actually quite flexible on formats, if you don't want to follow the standard one. (Although this flexibility isn't obvious from the information you are sent!) For example, at one regional meeting I organized a special session that met at lunch time, with two speakers each lunch hour. I checked that people would be able to buy sandwiches nearby and bring them into the room, and got approval from the secretary for that region, and then just did it.

I've never been involved with one, but there are local arrangements committees for meetings, and I suppose that volunteers for those are appreciated.

As to assorted communications of the AMS, most have their own editorial boards, and to work your way on to one of them you should start communicating your ideas to the current boards.

Sometimes people feel that they should not "burden" younger people with committee assignments; I think that people should be able to make that decision themselves, and that those who want to get involved should be welcomed with open arms.

Participation in the MAA
by Ken Ross

As with the AMS, there are many ways for people to participate in the MAA. At the national level, there are over one hundred committees. A list can be found in the Mathematical Sciences Professional Directory or on the web at www.maa.org. Some of the committees have specialized tasks, but many others involve various aspects of education or professional development for which young mathematicians would be very appropriate members. One of the strengths of the MAA is the very broad participation of its members, and new participants are essential to maintain its vitality. Therefore, volunteers are encouraged, though the MAA Committee on Committees is not able to appoint them all because many constraints need to be taken into account. Interested persons should contact the MAA Secretary indicating their interests and any relevant special skills or experience.

The MAA is much more than a national organization. There are twenty-nine semi-autonomous, lively sections that are defined geographically and cover the United States and Canada. The sections vary, of course, but all of them welcome new participants. All of them have officers with various tasks, including public information officers, contest coordinators and coordinators of student chapters. Some of the sections also have their own committees that deal with issues at the local level. Contact the section secretary to find out how you can get involved in the section. The list of section officers is also available in the Mathematical Sciences Professional Directory and on the web.

Beating the Cost of Meetings
by Frank Sottile

We all recognize the value in attending mathematics meetings. Unfortunately, attending these meetings is often costly, and you may not have access to travel money. I see four main expenses: registration fees, travel, food, and lodging.

Registration fees are often either negligible or may be waived if you write an organizer. An exception for this is a large meeting of a professional society, such as the AMS. In any event, they are often fixed. The cost of travel is also largely fixed, although some airlines offer student discounts. Some meetings are inconveniently located and you may feel a need to rent a car. Here, you can often rely upon the goodwill of others, and organizers often arrange some (often inadequate or inconvenient) local transportation. It is hard to control the cost of food, since much important socializing occurs at meals, and this is one of the main reasons for attending a conference. When not doing such socializing, I don't eat at a restaurant. A sandwich counter or fast-food stand is an acceptable alternative when I am with other young people. When eating alone, I prefer to graze at a grocery store (if available).

Lodging is perhaps the biggest expense under your control. Big meetings are often in fairly expensive hotels. My room at a conference is primarily a place to sleep and shower, and there are usually cheaper places than the conference hotel. At the 1994 winter meeting of the Canadian Mathematical Society in Montreal, I stayed in the Montreal youth hostel for $18/night. The dormitory accommodations were fine for me, and I saved quite a bit as the hotel was $100 ($50 if I had found someone to share a room). Besides the Youth hostel, there was also a YMCA at $30 for a single room.

While the Montreal hostel was two subway stops from the conference hotel, for the 1995 Joint winter meetings, a San Francisco youth hostel was across the street from the Hilton. The $34 room that I shared with my wife was adequate for our needs, and about 1/2 the price of the cheapest AMS alternative. While I did not investigate, there were other places intermediate between the Youth hostel and the hotels offered through the AMS in both price and comfort. These could be arranged by phone, or (now, in 1998) via the Internet. While such Spartan accommodations are not for everyone, not everyone needs luxury accommodations at a conference.

If anyone has a suggestion or tip on beating the cost of attending meetings, feel free to share it with the readers of this newsletter.

[1998 update: It is worth noting that since this article appeared, the AMS has begun listing cheaper alternative accommodations for the Joint meetings.]

Math Talks
by Curtis D. Bennett and Frank Sottile

Like many of you, I just returned from the joint meetings where I saw many talks of varying quality. This has prompted me to write an article on giving a good math talk. While most of the suggestions are obvious, I have seen many speakers who could benefit from these suggestions.

When preparing a talk, you should first ascertain who you will be speaking to. Then you should decide the objective of your talk. A guiding principle is that while everyone in the audience has chosen to attend, it is your job to make it worth their while to listen.

There are three basics which go into the preparation of a good talk: research, materials, and organization. Research is what most of us are familiar with. By this I mean knowing your topic well enough that you can answer questions honestly, whatever your audience. Specifically you should know some of the basic theory underlying the topic as well as some of the consequences of the results discussed. Ideally you should also know some of the history of the subject. Depending on the setting, your talk need not be original material. If not, then know the history well enough to give proper credit and references.

The materials for the talk are the handouts, the overhead slides, the chalkboard, or whatever else you need. You should have practice using the materials. Typically you give handouts when conducting a series of seminar talks on a topic. Handouts will help attendees keep track of what you are doing and give them something to refer to later. If you make handouts, be sure that they are neat and clear. Remember to make large headings for topics so that people can use your handouts easily.

Exercise particular care when using overheads. Clear, concise slides can hold your audience's attention, but messy, disorganized slides may be the swiftest way to lose it. You should use 18 point (1/4") type or larger if typeset and half an inch if handwritten (neatly!). Consider your slides carefully. When a slide goes on the screen, everyone will start reading it. Thus, don't put it up until you want it read. As well, keep the slides concise and to the point. You can fill in details verbally.

Board technique is as important in talks as it is when teaching a class. This means write neatly and think about what will be on the board at all times. If you will want something later in the talk, write it where you can save it without too much effort. Some of these touches can turn a good talk into a great talk. Similarly, bad board technique can take a good talk and make it horrible. Steven Krantz's book *How to Teach Mathematics* has several pages on blackboard technique. Every suggestion he makes is as appropriate for a lecture or seminar as it is for a classroom.

When organizing a talk, remember that not all of your audience will be paying attention at any given time. Try to make it easy for those who have drifted to rejoin you. You will also hold more of your audience for a longer time if you sufficiently motivate the subject early in the talk. Put important comments in the beginning, as well. Every good talk I have ever attended has been well-motivated from the start. Remember that it is hard work to watch a talk. As the speaker, you must continually convince your audience that it is worth their while to pay attention.

Try to give basic examples illustrating the main points of your talk. This can be very hard as basic examples sometimes don't exist. In that case, try and come up with examples that motivate the ideas even if they are not precisely examples of what you are speaking about. (In this case, DO warn your audience what you have done). Remember, your job is to help your audience understand the subject. They will be able to do so more easily if they have examples to ponder.

Carefully consider which details to include. While details are frequently at the crux of the matter in mathematics, they may also serve to confuse. Those having little to do with the ideas of the subject are best left out. I have been told that many details are best done "in private, between consenting adults." Always keep in mind the goal of conveying the ideas behind the subject, and let that be your guide.

Another organizational issue is one of time. Do not go over your allotted time. Nobody enjoys having a talk go overtime, and nobody minds when a talk is a little bit short. Ideally, have something at the end that can be omitted if necessary or lengthened if necessary. This will help you end your talk before the allotted time.

If people ask questions during your talk, give them honest, thoughtful answers. The audience usually asks questions to help them understand. If you don't know an answer, say so. If you have a conjecture, explain why you believe it. If, however, the issue brought up is subtle, then say so and mention what makes it subtle without going into too much detail. Little else annoys people more than speakers who pretend to understand more than they really do.

There are several common formats for talks. Colloquium talks are given to general audiences and should not do too much. In a colloquium talk, give some history of the field and motivate the problem. By all means state your main theorem and discuss the proof, but be sparing with the unimportant details. A colloquium audience is much more interested in understanding the field and the importance of the result than the details of the proof.

A second common format for a talk (or series of talks) is a research seminar. At your home institution, your audience is the (other) faculty, and your purpose is to keep your colleagues abreast of your work. When visiting another institution, your intended audience is the three or four specialists in attendance. They invited you to see what you are doing and to learn something of value. Make it worth their while. Having said that, this is not a carte blanche to bamboozle the rest of the people in attendance.

A third common forum for a talk is at a meeting of specialists in your area. Such talks are typically 10 to 20 minutes. These talks are often seen as "advertisements" for your research. Again you should be clear about the motivation of the question, but for an audience of experts, you can motivate the problem while assuming the audience knows a lot about the field. By their very nature, 20-minute talks skip details. As before, you need to figure out what is important in your work and explain that. The statements of the theorems are far more important than the proofs. The general idea behind the proof should be given if possible, but don't go through messy calculations if you can avoid it. Probably the best comment is that you should be willing to cut material.

A last format is the job talk. This is like a colloquium, but much harder to give. In this case, you MUST know at least a week in advance who your audience is and what they expect out of your talk. What makes this type of talk much harder to give is that members of the audience will be looking for different qualities. At a research school, many will want to see your research and they will want to know what makes it good. Some will also be interested in seeing how well you organize a talk. Others will be interested in how well you answer questions, and some people will have other interests still. At a smaller school, you might be giving a talk to undergraduates. At some schools, you might even be teaching a class. It is extremely important to do a good job at this talk. Many members of the department will base their final decision on the talks given.

In all cases, you should practice your talk before giving it. The amount of practice depends on the type of talk, and as a realist, I know that not every talk gets practiced. The best practice comes before a live audience (if you can find one). Grab some friends and ask them to watch you practice. Then listen to their comments. Practice helps you fine tune your timing, and it also helps you discover where the rough spots are. Lastly, practice gets you used to the materials of the talk. I think a job talk should be practiced a minimum of 3 times. That way, you avoid major gaffes, know what you are going to say, and can be ready for the unexpected. For other talks, one practice run is probably enough, although if you encounter difficulties the first time, it is probably worth it to practice again.

This advice should make it sound like preparing a good talk is hard work. While it may be hard work to prepare a good talk, there are many benefits. The obvious benefits are communicating a subject you love to your peers, and gaining a reputation for being a thoughtful and considerate speaker. In addition I find that preparing a talk is always a boon to my research. By seeking a way to express my ideas in a clear and concise manner, I am forced to rethink the basics, and this often leads to new, simplifying ideas. For this reason, talks (particularly research seminars) are good to give when you are in the midst of some research, not after you have tied up all of the loose ends. (Although in a job talk, I would rather the loose ends were tied up.)

Perhaps the best way to improve your speaking is by observing others and thinking about their presentation. When you are paying rapt attention to a speaker, consider why your attention has been captured. If you lose interest in a talk, think about what the speaker has done. If you drift back into a talk, what enabled you to do so? By emulating the good and discarding the poor, your speaking ability will surely improve.

It is also worthwhile to read about giving talks. One good reference is Paul Halmos' article on the subject, "How to Talk Mathematics" [1].

Further on Math Talks
by Ken Ross

I respond to the fine article above. I would add only one suggestion. First decide on the main results that you want to convey. Then write the talk in reverse order providing only those definitions and remarks that are needed to make the results clear. As an example, I used to give talks on lacunary Fourier series. I could talk for a whole hour about interesting results and interconnections without ever mentioning measures or Fourier-Stieltjes transforms, even though one couldn't understand the proof of the first basic result in the field without these concepts.

Organizing a Special Session
by Curtis D. Bennett and Frank Sottile

Last spring there was some discussion amongst the YMN Board concerning AMS special sessions and how speakers are chosen. The purpose of this article is to summarize this discussion and address the questions raised. Much of the information included below is in the *Manual for Organizing Special Sessions* which the AMS sends out, appropriately enough, to organizers of special sessions.

Sessions at AMS Meetings. The scientific program at AMS meetings has several components, typically invited addresses, special sessions, and contributed paper sessions. There is a real distinction between special sessions and contributed paper sessions. Every AMS member has the right to present a report of their research at AMS meetings, with obvious controls (no proofs that the Earth is flat). The contributed paper sessions provide such a forum. In a contributed paper session consisting of ten-minute talks (humorously referred to as speed math sessions), such presentations are grouped together under the title of some sub discipline of mathematics, *e.g.*, Functional Analysis or Graph Theory and Combinatorics.

We contrast these contributed paper sessions with the special sessions: the latter are often organized around the invited addresses; the speakers are invited or at least their talks are refereed; and the special sessions usually have a rather tight focus.

How the special session topics and organizers are chosen. Special sessions are essentially small research conferences that are piggybacked onto the AMS meetings. Sessions are usually organized around a single topic, often related to one of the invited addresses.

Anyone can propose to organize a special session at an AMS meeting. Deadlines and other such details are published in the Notices. Theoretically the selection of the organizers (or chairpersons) and the topics of special sessions is the responsibility of the Committee to Select Hour Speakers (of the AMS) for the meeting in question. In practice, however, the job is done by the associate secretary in charge of the meeting. If volunteers come forward, usually the secretary is quite happy to draft them.

According to one associate secretary, special sessions at section meetings are pretty much guaranteed to be accepted unless there is a conflict with an existing session. Furthermore it is not necessary to stick to one's own geographical section either. The January meetings, on the other hand, are a little more selective about special sessions. This is in part because there are far more proposals for special sessions and in part because of practical concerns about the number of sessions the meetings can accommodate.

Organizing a session can be a valuable experience and a pain in the neck. In many respects it is easier than organizing a one-day or weekend conference; many details such as room availability, advertising, and the special needs of participants are done by the AMS staff. Also, since the AMS schedule calls for early organization, there are often few(er) last minute details which demand your attention. This means that the organizer can manage to attend most talks, often an impossibility at other conferences. Also, while there is no travel money to offer speakers, most mathematicians are aware of this and even though the meeting may be off of the beaten track, many will try to attend anyway.

Some of the headaches of organizing a session concern making sure the speakers turn in their abstracts on time. Since the AMS requires abstracts about two and a half months in advance of the meeting and many speakers can only guess what results they may have to talk about that far ahead, many speakers will wait until the last moment to submit an abstract. As organizer, the job falls to you to remind speakers about the due dates for their abstracts.

Another duty that falls to the organizer is choosing speakers and answering requests to be included in the conference. The positive side is that you can be sure that the people you really want to have speak get invited. On the other hand, you may find that there are more people that you want to invite (or who want to speak) than there are time slots. This can be uncomfortable, although almost everybody is extremely understanding about this.

How the special session speakers are selected. The AMS Manual for Organizers lists five basic ways in which papers are selected.
1. The speaker is invited by the organizer.
2. The speaker volunteers by submitting an abstract three weeks earlier than the final abstract deadline and requesting that it be considered by the organizer for inclusion in the session.
3. The speaker volunteers by writing directly to the organizer.
4. The organizer asks to see all abstracts with a given two-digit classification number and selects certain ones for the special session.
5. After receiving all the abstracts, the associate secretary suggests by phone to the organizer that he or she might wish to include one or two particularly appropriate papers in his or her session.

According to the AMS Manual, about 80% of the papers selected at all special sessions are chosen by method (1). About 5% of the papers are selected by each of the other methods. In fact it is not unusual for special session organizers to know exactly who they want before the session is even announced.

Some of the editors of Concerns have found (3) to be a useful route to getting invited, when used in moderation. It is perfectly reasonable for the organizer to tell a volunteer that there is no room for them. It is also possible for the organizer to find themselves in a place where they will happily consider volunteers once they have heard back from several of the invitees. Often an organizer will invite people with the expectation that some will say no. Then, after a first wave, they may have more people to invite or they may wait to see who sees the announcement and wants to attend. If you are still a graduate student or if you only recently received your degree, you may wish to ask your advisor to approach the organizers.

In general we would always recommend getting in touch with the organizers if you are interested in talking. Unfortunately organizers will also have to say no occasionally because your talk won't really fit the session or because of time constraints. Ideally the speakers at a session will span several mathematical generations, and even if you are unable to speak, there is a great deal to be gained by attending a special session in a field you are interested in. After all, the point of the meeting and the session is to bring people together to discuss mathematics.

Some caveats about special sessions: For some reason (perhaps cultural) some fields of mathematics rarely have special sessions at AMS meetings. Other fields utilize this vehicle for many of their meetings. For this reason, the information in this article might not be as useful for some. The AMS is frequently looking for organizers of special sessions. So if you want to see a special session in your area, you should consider organizing one.

References

1. P. Halmos, *How to Talk Mathematics*, Notices of the AMS **21** (1974), no. 3, 155–158.
2. P. Halmos, *I Want to be a Mathematician: an automathography in three parts*, MAA, Washington, D.C., 1985.

Epilogue

A Pep-Talk

The majority of this book is devoted to describing what you *can* do during your first years after your dissertation, and so we conclude with a brief exhortation about what we, the editors, believe you *should* do:

Do what you love, and do it with enthusiasm.

Mathematics is a profession that allows you to develop your own niche and to excel there. You are fortunate; you belong to a mathematical community which is small enough that it is indeed a community—one you can turn to with questions and requests—and large enough that it has room for a wide variety of talents and tastes. This book is proof, if merely "proof by example", that individual mathematicians make noteworthy contributions, even early in their careers, in every aspect of our profession.

Take a chance, ask a lot of questions, go to meetings. And keep in touch.

A Selection of Professional Organizations and Agencies

American Mathematical Society, www.ams.org; (401-455-4000)
Mathematical Association of America, www.maa.org; (202-387-5200)
Society for Industrial and Applied Mathematics, www.siam.org;
   (215-386-9800)
American Mathematical Association of Two-Year Colleges,
   www.amatyc.org; (901-383-4643)
Association for Women in Mathematics, www.awm-math.org;
   (301-405-7892)
National Association of Mathematics,
   jewel.morgan.edu/~nam; (919-335-3326)
National Council for Teachers of Mathematics, www.nctm.org;
   (703-620-9840)
Young Mathematicians Network, www.youngmath.org;
   (subscribe at majordomo@youngmath.org)
Casualty Actuarial Society, www.casact.org; (703-276-3100)
Society of Actuaries, www.soa.org; (847-706-3500)
National Science Foundation, www.nsf.gov;
   (Director of Mathematical Sciences Division: 703-306-1870)

List of Authors

Below is a list of our authors, their current affiliations (and, if different, the institutions at which they wrote the articles in this book.)

Wendy Alexander (formerly Brunzie), Mills College (University of Montana)

Stan Benkoski, Daniel Wagner and Associates

Curtis Bennett, Bowling Green State University

Kevin Charlwood, Washburn University (Bradley College)

Annalisa Crannell, Franklin & Marshall College

Tom Davis, SGI

Julian Fleron, Westfield State College

Raymond Grinnell, University of the West Indies

Evelyn Hart, Colgate University (Hope College)

Charles Holland, Bowling Green State University

Paul D. Humke, St. Olaf College

Steve Kennedy, Carleton College (St. Olaf College)

Steven Krantz, Washington University

Lew Lefton, University of New Orleans

Daniel Lieman, University of Missouri (MSRI)

Terri Lindquester, Rhodes College

John D. Lorch, Ball State University (Wingate College)

Dana Mackenzie, Freelance Writer (Kenyon College)

Kevin Madigan, CNA Insurance Companies (Northwestern)

Tim McNicholl, University of Dallas (Northern Virginia Community College)

Robert Molzon, University of Kentucky (National Science Foundation)

Mark Montgomery, Grinnell College

Margaret Murray, Virginia Technical Institute

Jim Phillips, The Boeing Company

Richard Phillips, Michigan State University

Irene Powell, Grinnell College

Ken Ross, University of Oregon

Michael Sand, Raytheon Systems Company (Hughes Aircraft Company)

Mary Shepherd, SUNY-College at Potsdam

Paul Shick, John Carroll University

Karen Singer-Cohen, Wellesley College (Johns Hopkins University)

Ronald Solomon, The Ohio State University

Frank Sottile, University of Wisconsin, Madison (University of Toronto; MSRI)

Tina Straley, Kennesaw State University

Jean Taylor, Rutgers University

Leonard VanWyk, James Madison University (Suffolk U.; Hope College)

Stan Wagon, Macalester College

Todd Wilson, California State University, Fresno
    (Carnegie Mellon; Research Inst. for Symbolic Computation, Linz, Austria)

Mark W. Winstead, (University of California, San Diego)

Index